AF784533

BIOCOMPUTATION AND BIOMEDICAL INFORMATICS

BIOCOMPUTATION AND BIOMEDICAL INFORMATICS

U. S. Raghavender

Delve
PUBLISHING

www.delvepublishing.com

Biocomputation and Biomedical Informatics

U. S. Raghavender

Delve Publishing

2010 Winston Park Drive,

2nd Floor

Oakville, ON L6H 5R7

Canada

www.delvepublishing.com

Tel: 001-289-291-7705
001-905-616-2116

Fax: 001-289-291-7601

Email: orders@arclereducation.com

© 2019 Delve Publishing

ISBN: 978-1-77361-394-9 (Hardcover)

This book contains information obtained from highly regarded resources. Reprinted material sources are indicated and copyright remains with the original owners. Copyright for images and other graphics remains with the original owners as indicated. A Wide variety of references are listed. Reasonable efforts have been made to publish reliable data. Authors or Editors or Publishers are not responsible for the accuracy of the information in the published chapters or consequences of their use. The publisher assumes no responsibility for any damage or grievance to the persons or property arising out of the use of any materials, instructions, methods or thoughts in the book. The authors or editors and the publisher have attempted to trace the copyright holders of all material reproduced in this publication and apologize to copyright holders if permission has not been obtained. If any copyright holder has not been acknowledged, please write to us so we may rectify.

Notice: Registered trademark of products or corporate names are used only for explanation and identification without intent of infringement.

Delve Publishing publishes wide variety of books and eBooks. For more information about Delve Publishing and its products, visit our website at www.delvepublishing.com.

DEDICATION

To all the great souls who contributed to knowledge of the world…

ABOUT THE AUTHOR

Dr. U. S. Raghavender did his Masters in Physics, with specialization in Condensed Matter Physics, followed by his doctoral thesis in X-ray Crystallography of Designed molecules. This was followed by post-doctoral stints in India and abroad, in both academic institutions and non-academic institutions. He has been working on computational approaches of biomolecules, since the past 12 years. He has authored book chapters, published research in internationally acclaimed journals. He also serves as a peer reviewer of research articles published in both national (Journal of Biosciences) and international journals (BMC Bioinformatics, Nucleic Acids Research etc.). He has a long term goal of addressing computational aspects of regulatory mechanisms in cell.

TABLE OF CONTENTS

PREFACE

Almost everybody is familiar with the catchwords like informatics, medicine, bioinformatics, and computation. Certainly, there should be a way where all these seemingly disparate fields will come together and contribute to the betterment of society. Computation has seeped into every nook and corner of the presently digitized world. Medicine and healthcare are making their presence more than ever before. Everyone seems to be aware of the diseases, symptoms, prognosis, and diagnosis without possessing a medical degree. Thanks to the internet age and the ever-increasing rate and pace at which the data are being shared across the internet.

Biomedicine is an amalgamation of many subdisciplines including biology, bioinformatics, medicine, and healthcare. Informatics, in a general sense of the world, corresponds to the way we encode data into computers and tools or programming skills which help us derive meaningful and insightful information from such data. The web is full of varied and interesting data repositories.

In this book, I attempt to introduce the broad field of biomedical informatics to the reader. We do not and cannot every aspect of the field in the present endeavor. But, we do cover in a sense some of the state-of-art topics in this book. We start with an introduction and move on to subsequent chapters in a style, which involves hands-on programming. We have employed such an approach in our previous books and found it to be more enlightening than a verbose and dry introduction of the concepts. As this book involves computation, there will be a substantial component of coding. The reader is expected to have a basic understanding of programming languages; Python being preferred.

1
CHAPTER

BIOMEDICAL INFORMATICS

CONTENTS

Biomedical Informatics is an interdisciplinary field seeking to study and advance the use of biomedical data for the scientific study, problem-solving, and charting the best path to improve human health.

According to the American Medical Informatics Association, biomedical informatics applies theories and processes to help generate, store, retrieve, use, and share biomedical data by advancing computing, communication and information science as it applies to biomedicine.

It investigates, simulates, experiments with and translates a wide swath of biological systems to connect basic and clinical research with practical application for the overall betterment of healthcare. It does not lose sight of the human element, incorporating social and behavioral science into the equation when evaluating technical solutions.

Biomedical informatics can be used by clinicians, scientists, researchers, educators and other professionals whose primary job function is to connect people with information and technology.

1.1. WHAT IS INFORMATICS?

We start this book, with a bird's eye view of Informatics. Basically, informatics stands for information and the science which explains it in all its aspects. This includes, data as its basic unit. With storage and processing making its two limbs. A basic aspect of informatics is communicating the results and analyzes. As of 2018, it can be linked with the information which is computerized. This field has expanded its wings to encompass medical informatics or health informatics.

Healthcare is one of the most dominant users of patient data. This is encoded in what is known as Electronic Health Records (EHR). As today's seemingly infinite amount of data continues growing, informatics is used to help process, analyze and communicate large chunks of data while seeking clues to support and sustain public health and enhance the quality of care in the field of biomedical informatics.

AMIA has proclaimed that Biomedical Informatics (BI) can play a vital role in assisting researching, and analyzing clinical care data. It can play a prime role in monitoring public health through data mining, language processing, cognitive science and human interface design.

BI is science, a subfield of computational science. There is a great emphasis on creating new algorithms. In addition, databases and extracting meaningful information plays a very important role in BI. The field spans clinical research, genomics, proteomics and public health data. On a more personal level, health informatics can gather, store and more quickly disseminate an individual's healthcare information between various providers.

1.2. A BRIEF HISTORY OF INFORMATICS

The term "informatics" was coined as the name of one of the first software companies. The founder of the company Walter Bauer authored a paper in 1996. The authors put forward their view on the choice of the word "informatics" was in the sense of "the science of information." Phillipe Dreyfus also coined the term, simultaneously, for a software company named SIA (Societe pour L'Informatique et Applique).

Let us now see what are the key areas in Informatics.

1.3. KEY AREAS OF INFORMATICS

In terms of practice, the AMIA has divided BI into five key areas:

- **Translational Bioinformatics** (TBI) – this involves ways to research, analyze and interpret medical data. This also involves developing ways to use and integrate such information into clinical and healthcare applications.
- **Clinical Research Informatics** (CRI) – this applies informatics concepts from clinical trials to reveal new discoveries in healthcare and disease.
- **Clinical Informatics** (CI) – this area involves application of informatics to the delivery of a range of healthcare services. This caters to the needs of all health professionals such as nurses, pharmacists and physicians.
- **Consumer Health Informatics** (CHI) – this area focuses on data that consumers can use to improve their own healthcare. This makes available to consumers the appropriate medical information and in a form useful to consumers.
- **Public Health Informatics** (PHI) – this involves application of informatics to areas of public health. Public health is a broad topic involving prevention, reporting and preparation of data for groups of people instead of individual consumers.

A common thread through all these is informatics related to human biology. Everything related to Biomedical Informatics can be grouped under a big umbrella of 'Bioinformatics.' But, the fields have diversified so much that each field has its own distinction and the overlaps are vanishing as we are seeing advancements in informatics and related concepts. With regard to careers, the diversity of bioinformatics applications has opened a range of informatics careers. This has expanded beyond healthcare industries. Employers in the pharmaceutical, software and biotechnology industries are seeking to hire trained professionals trained in informatics.

Without a much ado, let us venture into different subfields of BI. But, before that let us see what career options are open to individuals with good informatics skills.

1.4. INFORMATICS CAREER OPTIONS

We list a few options below, although there could be many more opportunities:

- Healthcare IT Project Manager: In this role one creates and supervises research projects. This also involves maintaining communication with other participants and suitably adapting to the changes as and when needed;
- Nursing Informatics Specialist: This involves trains nurses on protocols of record keeping. With special emphasis on enhanced accuracy of records, which helps in data integration into patient care;
- Chief Medical Information Officer (CMIO): In this role, which is quite advanced and requires a lot of experience, one usually oversees information processing, IT systems and improvement of information technology infrastructure. It may optionally involve designing software applications, as well as conducting training for beginners;
- Health Informatics Consultant: As a consultant, one would offer advice about facilities with regard to meeting federal mandates, software installation, troubleshooting and training.
- Health Informatics Director: A senior level executive position, essentially involving coordinating data use across different departments. It may also involve handling training and interfacing with medical personnel on varied issues, including technological.

With the data deluge in every field, particularly in big pharma and biotech, software companies are clamoring to hire professionals with experience in bioinformatics. In addition, the hunt is for data analysts and professionals who are adept at identification, compilation, analysis, and visualization of huge amounts of biological and healthcare information.

1.5. BIOMEDICAL INFORMATICS SUB-FIELDS

As pointed out in Section 1.3, Biomedical informatics has many subfields. Almost every subfield in BI can be further grouped into four subfields. These four subfields are: public health, clinical, imaging, and bioinformatics. This would also mean a concomitant incorporation of the component scientific disciplines, which include computer sciences, clinical sciences, basic biomedical sciences, cognitive sciences, bioengineering, management, epidemiology and statistics. In a nutshell, it is a field with participation of professionals from varied scientific disciplines. Let us now see what the four subfields are:

- **Public health Informatics** (PHI). The primary focus of PHI is not individual-centric, but encompasses the whole of society. This involves preparing a large section of society for internal and external threats such as a bioterrorist attack. It also involves expanding upon an existing application such as the National Notifiable Disease Surveillance System.
- **Clinical informatics.** This is a field which is patient-centric. It involves collecting, managing, storing and disseminating electronic medical records for each patient.
- **Imaging informatics.** This is a specialized area which focuses on tissues and organs of patients. This primarily involves advanced imaging applications like computerized tomography scanner, which utilizes software algorithms to render a three-dimensional image of specific organs.
- **Bioinformatics.** This is a well-established field and needs no explanation. Briefly, it deals with computational methods to read, parse and encode molecular and cellular processes to derive meaningful information.

1.6. WORKING IN THE FIELD

With a brief overview provided in the previous sections, the natural question which arises in the mind of a curious reader is what it would be like to work in a field like this. Certainly, the novice need to first learn the history of the discipline. A great emphasis is on understanding data at your disposal. Hence, a good deal of programming skills is a must for the entrant. In addition, the ability to analyze the problems requires a good understanding basic concepts in statistics and allied disciplines. Producing viable solutions is an integral part of the entire processes and the professional is also to provide a reason as to why that solution is better than a competing option.

The informaticians must be capable of carrying out a solution to assess its potency while improving its design, if necessary. There is a great emphasis on innovation, as in any other field. It is a key for creating new technologies. This would mean collectively working with peers from other disciplines. Finally, biomedical informaticians must be able to communicate findings effectively, both in writing and orally. This is a very important aspect of public dissemination of information and analyzes. Hence, the informaticist is required to develop skills in the area of presentations which involves dashboards and story telling. What I mean is that the professional needs to hone skills in the area of data visualization and interactive visualization.

All the above require a basic ability to understand biomedical concepts, including health-related information, and their models to molecular details for the entire human population.

1.7. MANAGEMENT IN HEALTH INDUSTRY

We are already seeing a transition from handwritten medical records to electronic medical records, project managers are playing a very important role in this transition. In fact, the role of project manager becomes very critical as one who plays an important role in healthcare industry and who also enjoys working with technology. The adept may find a comfortable career working as an IT project manager in healthcare informatics.

With the federal law dictating the move from paper to electronic records, the need for hiring skilled project managers is on the rise. Certainly, the job requires substantial experience at an executive level still the lucrative profession requires proper education and proficiency in both project management and health informatics.

1.8. IT PROJECT MANAGERS

Let us get a peek at what project managers are engaged in Biomedical or Healthcare Informatics. In IT project managers typically oversee large, often complicated projects that involve technology. The specifics differ based on the industry, but generally, project managers aim to produce quality results with maximum efficiency, on time and on budget. In the healthcare industry today, the largest IT project for most facilities is converting patient records from paper to electronic databases. Federal mandates require healthcare providers to make the shift to electronic health and medical records or face serious financial penalties with regard to Medicare and Medicaid reimbursement.

Basically, project managers help improve business processes to ensure consistent, measurable results. The job in healthcare industry may look a little dry and monotonous as it involves managing records. Those involved with health informatics must manage records in such a way that the most up-to-date information is readily available to doctors, specialists, nurses and patients, so that better healthcare decisions can be made. This reduces costs and improves patient outcomes.

1.9. CAREER OUTLOOK FOR HEALTHCARE IT PROJECT MANAGERS

There is a perception amongst experts, that the number of healthcare IT project manager jobs will grow considerably in the years to come. The present federal mandate promotes transition from paper patient records to electronic records. All this means that one should be aware of software systems, their installation and management.

1.10. HEALTHCARE IT PROJECT MANAGER EDUCATION

Obviously, one needs to have qualifications in Business Administration. One needs to complement it with a certification in Project management. This can be achieved by working with or under senior level managers in healthcare industry. One can get a good understanding of the field and scope of the subject by applying for a Master's degree in Health Informatics or undergo a graduate certification.

1.11. PUBLIC HEALTH INFORMATICS

Let us take another subfield (PHI) and discuss it to some practical extent. Essentially, it is a confluence of two distinct disciplines. Information sciences and public healthcare practice. This is an ever-expanding field and the methods are evolving to accommodate the diverse varieties of data in healthcare industries. Basically, it is an avalanche of data from health repositories. We are aiming to efficiently manage and disseminate information from public health repositories using IT.

Public health informatics is more than just looking, managing and cataloging existing information. When looked at analytically there is a good opportunity to reimplement old ventures using new methodologies. Like Medical Informatics, PHI aims at employing information science and technology to help in improving health of public at large. The similarities end here and the differences emerge beyond IT. This was pointed in a report from U.S. Centers for Disease Control and Prevention.

So, the question is what is the primary focus of public health informatics?

The gamut of the application of information science technology is to promote the health of all populations, and not just a few specific individuals. PHI essentially uses different applications to prevent disease and injury for populations, i.e., a large group of people, at risk.

Public health informatics utilizes applications that prevent disease and possible injury by changing the environment or tweaking the conditions that put large groups of people at risk. It does not limit itself to particular contexts such as social, behavioral or environmental; instead, public health informatics considers how to effect prevention at all vulnerable points in the path to disease, disability or injury. And it takes into consideration the governmental context through which public health is practiced.

Because it focuses on the overall population, as opposed to the individual, public health informatics must utilize data from a myriad of sources – including hospitals, law enforcement and social services – while incorporating results from surveys, inspections and more.

As technology continues to evolve, and healthcare continues to expand, public health informatics has become a vital component, helping providers and government agencies address the changing times and prepare for new and unexpected threats such as antibiotic-resistant infections and chemical and biological attacks. Such an attack would require immediate detection and analysis of the agent in order for public health officials to address the threat and respond appropriately to prevent widespread human harm. That is why informatics training is essential to help prevent system failures and to allow for knowledgeable and competent leadership. Public health informaticians must understand both disciplines in order to develop and utilize information technology that doesn't impede on privacy and confidentiality.

1.12. PUBLIC HEALTH MANAGEMENT

The changing landscape of healthcare is creating a huge demand for health data analytics for Population Health Management (PHM). PHM involves using healthcare analytics to help establish best practices related to the treatment, prevention and self-care of patients throughout a set population. As medical recordkeeping moves from paper to digital, the ability to share information critical to patient care and overall health outcomes is broadening.

1.13. UNDERSTANDING POPULATION HEALTH MANAGEMENT

PHM isn't a new concept, but it's one that's gaining more ground as electronic health records (EHR) makes mining data to improve patient care more feasible. Analytics may also be used to reduce healthcare costs while also cutting fraud, abuse and waste in the $3.0 trillion healthcare industry in an effort to maximize investments and improve results.

PHM has a few distinct goals that are achieved through the use of data mining. The overall goal is to use analytics to determine what works to improve patient outcomes and/or reduce costs while identifying what changes need to be made. The benefits of the in-depth analysis associated with PHM are valid for not only patients, but also providers, policymakers, employers and insurance companies.

Though effective PHM has long been a goal, realization has been hampered by an inability to mine all data related to populations and population subsets. The practice has been limited to public health clinicians simply because of the limiting factors associated with healthcare information sharing prior to the implementation of EHRs. For example, though a hospital might have a large data pool of information about diabetics in a certain geographical area, the picture of their medical status isn't fully complete without information also gleaned from local urgent care centers, private medical practices, pharmacies and other medical stakeholders. EHRs are making the collaboration and information sharing required by PHM more feasible. The practice in this field may require one to:

- Identify patient populations that might require targeted care
- Examine characteristics of the subpopulations
- Create reminders for patients that are actionable
- Track performance
- Make data more accessible for improved analysis

1.14. WHERE BIG DATA COMES INTO PLAY

Big Data (BD) is a term employed to data in huge amounts, being continuously generated, and their storage and communication across varied networks. This also involves the processing of distributed computing. In the case of biomedicine, big data involves the implementation of EHRs and other forms of scalable, streaming data that can be used to help with real-time decision-making related to patient care and PHM. As more and more medical practices move to EHRs, as per the Federal directives, the ability to mine data outside of a hospital's boundaries is becoming feasible. That means analysis of more complete, integrated data is also becoming possible.

Let us take a simple example to illustrate how big data comes into play in healthcare. Let us consider a population in a particular region who have had some form of cardiac problem. Some of the testing records may be available within the hospital; some other patients' records, in a silo, may be held within a testing facility. The availability of EHR's is making it possible to get a better picture of patient status. The technology is also making it possible in obtaining a real-time information that prevents duplication of services and promotes improved quality of care. Although we have mentioned one use case of big data, but the ultimate goal is to be able to use data across patient populations to improve care and lower costs. Widespread use of big data is not yet available in PHI. As more practices move towards EHRs, the ability to aggregate data, filter it and analyze it is becoming a real possibility that could have a very positive impact on patient outcomes as PHM becomes more attainable.

1.15. WHAT ARE EHR'S?

Electronic health records (EHRs) are detailed accounts of health information, generated by a patient's healthcare providers across the industry. An

EHR typically contains information such as the patient's demographics, contact information, vital signs, allergies, medical history, current and past medications, immunizations, radiology reports and laboratory data.

EHRs are designed to be shared across the spectrum of a patient's providers to help deliver a more comprehensive health experience with the most accurate information possible. Today, a number of healthcare providers have found that with the implementation of electronic health records also comes with decreased medical costs and improved patient care.

1.16. ELECTRONIC HEALTH RECORDS: THE PAST, PRESENT, AND FUTURE

The evolution of EHRs started back in the late 1960s when multiple efforts in support of EHRs were appearing throughout the U.S. It all began with new technology that changed the way we record and share data, and healthcare providers wasted no time in seizing this opportunity to improve the delivery of healthcare. Three problems were identified:

(1) to eliminate the logistical problems of the paper records by making clinical data immediately available to authorized users. This would essentially stop unavailable or undecipherable clinical records;

(2) to reduce the work of clinical bookkeeping required to manage patients – no more missed diagnoses when laboratory evidence shouts its existence;

(3) to make the informational 'gold' in the medical record accessible to clinical, epidemiological, outcomes and management research.

1.17. WHERE WE ARE NOW WITH ELECTRONIC HEALTH RECORDS?

From 2001 to 2011, the percentage of physicians using EHRs grew from 18% to 57% and the percentage continued to increase in the years following. By the year 2015, 63% of healthcare providers had adopted EHRs. The percentage for 2016 went down slightly to 59% due to a larger sampling size. Looking at these percentages, it is clear that EHRs have come a long way over the past few years and should only continue to grow as time goes on.

Here are some EHR advancements to look for in 2018:

- **Cloud-based EHRs** – A number of industries, including healthcare, are making the move to software applications in the cloud. Healthcare providers are finding that it is much more cost effective to implement and update EHRs in the cloud because it removes the constant need for new hardware.
- **Improved Patient Portals** – Patients truly enjoy the convenience that comes with a patient portal. The patients can use them to do various things, including following up with their provider, paying medical bills or being notified of other kinds of critical information. Some pretty big upgrades to these portals are expected to come in 2018.
- **Growth of Telehealth** – The market for telemedicine is expected to exceed $30 billion by the end of the decade, according to healthcareitnews.com. With telehealth technology, doctors can do consults with patients prior to surgery without having them come into an office or hospital. This is especially convenient for members of the senior population who have a hard time leaving their homes to be evaluated by a doctor.
- **Mobile Friendly EHRs** – Doctors and nurses cannot always be tied to a computer screen. They need to be able to access EHRs on mobile devices, such as smartphones or tablets. In order to fill this need, many EHRs will have to update their interfaces to be mobile friendly.

1.18. THE FUTURE OF ELECTRONIC HEALTH RECORDS

Most healthcare providers have moved past the implementation phase of EHRs and are focusing on the continuous optimization of them. From this point on, all workflows related to a patient's care are expected to be handled by EHRs. These workflows include handling external documents, messages and general communication with patients. Patients should only need to log into one patient portal to access all of their physicians and communicate with them.

EHRs will also need to provide open access to internal data. Clinical quality measures, demand that will affect staffing levels and disease trends are all good examples of reports that healthcare providers will want to have access to. Essentially, the role of EHRs as repositories for patient data will slowly go away.

The future of EHRs is all about expanding the interconnection between patients and their healthcare providers. *Patients' health data should be connected across all of the places where care is sought.* Readily available information, as well as direct access to communication with any healthcare provider, should always be just a click or a swipe away.

Over the past several years, the U.S. healthcare industry has been undergoing major changes, one of which is the transition from paper to digital record keeping by hospitals, doctor's offices, clinics and nursing facilities. The federal government and health organizations have invested billions of dollars to fund the hardware, software and training necessary to complete the changeover.

The mandated switch to electronic records has garnered plenty of news coverage, with stories about electronic health records (EHR) and electronic medical records (EMR) peppering both medical and mainstream publications. The terms are often used interchangeably, which can lead to confusion.

However, there are distinct differences between EHR and EMR.

1.19 WHAT ARE ELECTRONIC MEDICAL RECORDS?

An electronic medical record (EMR) is a single practice's digital version of a patient's chart. An EMR contains the patient's medical history, diagnoses and treatments by a particular physician, nurse practitioner, specialist, dentist, surgeon or clinic.

EMRs offer several advantages over paper records:

- Better data tracking over time.
- Timely reminders for patient screenings and preventative checkups.
- Improved patient care.

1.20. WHAT ARE ELECTRONIC HEALTH RECORDS?

An electronic health record (EHR) is also a digital version of a patient chart, but it is a more inclusive snapshot of the patient's medical history. Electronic health records are designed to be shared with other providers, so authorized users may instantly access a patient's EHR from across different healthcare providers.

The benefits of EHRs include:

- Streamlined sharing of updated, real-time information.
- Access to tools that providers can use for decision-making.
- A complete medical history of the patient, from allergies and radiology images to lab results.

1.21. HOW DO ELECTRONIC MEDICAL RECORDS DIFFER FROM ELECTRONIC HEALTH RECORDS?

It's easy to remember the distinction between EMRs and EHRs, if you think about the term "medical" versus the term "health." An EMR is a narrower view of a patient's medical history, while an EHR is a more comprehensive report of the patient's overall health.

Here are a few more ways EMRs and EHRs differ:

- An EMR is mainly used by providers for diagnosis and treatment.
- EMRs are not designed to be shared outside the individual practice.
- EHRs are designed to share a patient's information with authorized providers and staff from more than one organization.
- EHRs allow a patient's medical information to move with them to specialists, labs, imaging facilities, emergency rooms and pharmacies, as well as across state lines.

1.22. 3D IMAGING IN HEALTHCARE DELIVERY

With the advent of 3D printing and imaging, healthcare and biomedicine has donned a new face. A prosthetic limb. Healthy tissue. A lifelike model of a liver. All from a printer. Already, the impact of 3D imaging is being felt in automotive and aerospace industries, among many others. Biomedicine and healthcare is no exception. The marvels of 3D imaging may seem like science fiction today, but they are proving to be very real and have the potential to be massively transformative in U.S. healthcare in the near future.

1.23. MEDICAL INFORMATICS

Medical informatics is found at the intersection of healthcare and technology. It is where skills in both medical and computer sciences come together in an effort to improve healthcare and patient outcomes. Professionals in this hybrid field draw on expertise from both disciplines to put technology to its best use in patient care, clinical and research settings.

1.24. TYPES OF WORK IN MEDICAL INFORMATICS

Medical informatics professionals are tasked with using information technology to its greatest advantage in the healthcare industry. This means they are responsible for such tasks as:

- Creating, maintaining or facilitating new ways for medical facilities and practices to keep electronic health records (EHR),
- Improving communication between healthcare providers and facilities to ensure the best patient outcomes,
- Storing, managing and analyzing data for research,
- Assisting with complex, technology-dependent research, such as that involved in human genome sequencing.

1.25. SPECIALTIES AND CAREERS IN THE FIELD OF MEDICAL INFORMATICS

Medical and computer science professionals who wish to transition into this field with a master's or doctoral degree and students heading straight into this arena will find a number of specialties exist. They include:

- **Bioinformatics**: Practitioners in this specialty are concerned with storing, retrieving, sharing and helping analyze biomedical information for research and/or patient care. Subspecialties include chemical, nursing and dental informatics.
- **Public health informatics**: This specialty involves the use of technology to guide how the public learns about health and healthcare while also ensuring access to the latest medical research. Professionals also ensure public health practices have access to the information they need.

- **Organizational informatics**: The focus here is ensuring a smooth flow of communication within a healthcare organization.
- **Social informatics**: These specialists study the social aspects of computer science while gaining insights into how information technology affects social environments and how social environments affect information technology.
- **Clinical informatics**: This is the application of informatics and information technology for clinical research and patient care. Professionals leverage information technology for medical education, patient education and students, among others.

Medical informatics job growth is predicted at about 21% through 2020, according to the U.S. Bureau of Labor Statistics. That agency doesn't break out potential earnings, but the American Health Information Management Association (AHIMA) estimates mid-range salaries for health informatics consultants in the upper $80,000s. Those in management circles can see salaries climb as high as $200,000. Those interested in medical informatics are urged to conduct their own research. Job availability and salaries may vary based on position, employer and geographical location.

As medicine and technology continue to intersect, bioinformatics is the science that is driving the future forward. This distinct discipline is a hybrid that draws on both biological and computer science expertise. As medicine and technology continue to intersect, bioinformatics is the science that is driving the future forward. This distinct discipline is a hybrid that draws on both biological and computer science expertise. Bioinformatics is a subsect of informatics that has far-reaching implications to be sure.

1.26. WHAT IS BIOINFORMATICS?

Bioinformatics involves the application of computer technology to manage volumes of biological information. Computers are used to not only store, but also gather, analyze and integrate biological and genetic data that can then be applied for such uses as gene-based drug discovery and development. The discipline grew out of the massive amounts of information accumulated by the Human Genome Project, which has required advanced computer technology to help sequence the entire human genome, which amounts to some 3 billion base pairs. Thanks to human genome research, applications for bioinformatics extend to a number of areas. Applications of bioinformatics include:

- **Basic research**: Bioinformatics is relied upon to assist with research into such areas as comparative and evolutionary genomics, functional genomics and genome-wide association analysis. The mass of genomic data generated by high-performance technologies has made this discipline vital for storing, managing and analyzing genomic information commonly used in biological research today.
- **Biomedicine**: The application of bioinformatics is proving quite useful in this field as the human genome has helped unlock the genetic components for many diseases. Potential applications include drug discovery, personalized medicine, preventative medicine and gene therapy.
- **Microbiology**: The potential applications here involve the study of microorganism genomes to assist with biotechnology developments, waste cleanup, climate change, antibiotic resistance and more.
- **Agriculture**: By sequencing the genomes of animals and plants, genetic knowledge can be gleaned to help produce stronger crops while improving the quality of livestock.

1.27. WORKING IN BIOINFORMATICS

As bioinformatics emerges as its own distinct field, there are a number of positions for those with proper qualifications which can be pursued. Some of the job titles in bioinformatics include:

- **Bioinformatics specialist**: These professionals are required to provide bioinformatics and data analyzes services, training on available tools and offering scientific writing assistance to researchers.
- **Bioinformatics research scientist**: These scientists apply information technology and computer science in the study of biology. They are involved in research, study, analysis and processing of genomic and other forms of biological information.
- **Bioinformatics analyst**: The job of these professionals includes analyzing, manipulating and interpreting large data sets for use in research.
- **Biology informatics software engineer**: Biology informatics engineers' are involved in creating or advancing software for

use in bioinformatics research. They may work in experimental design, help develop new tools to analyze and validate data, and seek ways to extend data acquisition capabilities.

1.28. CONCLUSIONS AND TRENDS IN 2018

As the second decade of 2000's is coming to a close, we can take a quick peek at the recent developments and trends in the area of healthcare and biomedical informatics. We expect to see some far-reaching and widely applicable advancements in healthcare in 2018.

1.28.1. Telemedicine and Mobile Health Become Commonplace

It's been said in past years, but this time telemedicine and mobile health really are having their biggest impact on healthcare. We've already seen significant growth in telehealth efforts, but in 2018, it's expected to grow toward ubiquity due to growth in the amount of data being shared across providers, and increased usage of electronic health records (EHRs) and mobile health devices by consumers.

Mobile health technology is at the center of patient engagement strategies, a key component in the value-based healthcare movement. New processes for the vetting and review of mobile apps may push their use to new heights.

1.28.2. Deep Learning in Healthcare Informatics

Deep learning has in recent years set an exciting new trend in machine learning. The theoretical foundations of deep learning are well rooted in the classical neural network (NN) literature. But different to more traditional use of NNs, deep learning accounts for the use of many hidden neurons and layers—typically more than two—as an architectural advantage combined with new training paradigms. While resorting to many neurons allows an extensive coverage of the raw data at hand, the layer-by-layer pipeline of nonlinear combination of their outputs generates a lower dimensional projection of the input space. Every lower-dimensional projection corresponds to a higher perceptual level. Provided that the network is optimally weighted, it results in an effective high-level abstraction of the raw data or images. This high level of abstraction renders an automatic feature set, which otherwise would have required handcrafted features.

2
CHAPTER

PYTHON: AN INTRODUCTION TO SCIENTIFIC COMPUTING

CONTENTS

2.1. THE ROLE OF COMPUTING IN SCIENCE AND INFORMATICS

Traditionally, science has been divided into experimental and theoretical disciplines. During the past few decades, computation has played an important role, rather very important, in science and many of its sub-disciplines. The field of scientific computing has had its inception for addressing the following:

A way to bridge the theory with computing

This is the reason why novice, and at times advanced users, often feel that computing complements scientific theory. But, the field of scientific computing has evolved beyond this obvious definition. It now aims to complement experimental work. In this sense, it is often viewed as a new branch of science by many scientists. The opinion is divided, as experimentalists are often of the view that computation should subserve their observations. The indispensible complementarity of scientific computing in many diverse areas of science is evident when one reads many technical scientific papers or even in popular magazines (like Scientific American). Numerical calculations and computational modeling has made its way into the lives of every bench scientist.

It is an ethical practice in scientific field to make available different methods employed to make a certain claim to the public at large. For example, in theoretical sciences, derivations, proofs and other results are published in full detail, or made available upon request. If an author does not provide such a proof, that work is considered of questionable credibility. Technically, it would hinder other people to reproduce results. Reproducible research is one of the primary tenets of modern scientific research. This is all the more important for scientific computing.

In computational sciences, now-a-days the trend is to share the computational code by depositing code at their respective journal websites as supplementary materials to scientific papers. As a complement, the authors are encouraged to put the code for public display at the author's own website. Presently, the trend is to put the code on *github*. This is open source by default. Hence, accessible to one and all free of cost. This has greatly enhanced the reproducibility of scientific observations.

A recent trend which has been highly promoted amongst scientific circles is to share the methodology and scientific (computational) protocol employed in research in the form of *notebooks*. It is highly encouraged that the reader gets acquainted with Jupyter Notebooks (previously IPython notebooks). Most of the code in this book can be reproduced with such notebooks. In R environment, we have R markdown. But, nowadays Jupyter notebook also supports R kernel amongst support for many other languages. Hence, in a single notebook one can work with several languages to arrive at a meaningful research analysis.

2.2. REQUIREMENTS ON SCIENTIFIC COMPUTING

Any scientific enquiry or conclusion has two prime cornerstones – *Replication* and *Reproducibility*. Let us look at the implications of these two important concepts in the context of scientific computation:

- **Replication**: Replication does not mean that many copies of the same method are produced. It means that the work (code and examples) can be downloaded and worked upon in an independent environment. It should also lead to results in any environment, preferably identical environments. For example, an author of a scientific paper which has a substantial component of numerical calculations should be able to rerun the simulations. It involves not only rerunning but also replicating the results upon request by peers or other fellow scientists. By the same token, fellow scientists should also be able to perform the same calculations and obtain the identical results. Hence, the need to provide the information about the methods used in a certain research in publications.
- **Reproducibility**: The results obtained from computational studies must be reproducible. This could be with the same method, or a similar method. At times, reproducibility should be ascertained with a completely different method.

Summarizing, a sound scientific result must be reproducible and replicable.

What do we do to achieve such an objective is an obvious question. We can take the following approach:

- **Notes**: Primarily, try to keep and take regular notes of the exact source code. Also, keep track of the version of software and data that was used for analysis and figures in published papers.

- **Versions**: A regular record of correct version of external software that was used for research needs to be kept. It is also advised to keep access to the exact environment used to carry out research. The above two points can be mentioned as comments in the code itself, for ready reference.
- **Backup**: It is a good practice to preserve old code with substantial comments for future reference. This not only increases future reproducibility but also serves as a good reference.
- **Credibility**: As a professional, you should be able to communicate the results with any interested person anytime. Hence, you should meet the request with a smiling face and come forward to provide help.
- **Online publishing**: Publish your code to online places like github, for ready accessibility to other scientists.

2.3. PYTHON

By now, everybody knows what Python is in computational field. Certainly, the grip of Python is tightening as days are passing (pun intended). Let us see what Python is in simpler terms. In a layman's language it is a modern, general-purpose, object-oriented, high-level programming language. What makes such a popular language, is answered by observing the general characteristics of Python (basically two important features):

- **Clean and Simple Language:** Python is easy-to-read and the coding is intuitive. In addition, it is easy-to-learn with minimalistic syntax. Very easy to maintain, scales well with varied size of projects.
- **Expressive Language:** Most importantly, it requires very few lines of code, fewer bugs and easier to maintain. Debugging is also very easy.

Let us now see the technical details, which characterize Python (basically three):

- **Dynamically Typed:** In Python, there is no need to define the variable types, function arguments or return types.
- **Automatic Memory Management:** Allocation and deallocation of memory for variables and data arrays is automatic in Python and the issue of memory leak bugs is minimal to null.

- **Interpreted:** No need to compile the code. The Python interpreter reads and executes the python code directly.

Let us look at some of the advantages and disadvantages of Python. First let us see the good side.

- The primary reason for the popularity of Python is the ease of programming. This substantially minimizes the time needed to develop, debug and maintain the code.
- Although borrowing from C++, Python is really well-designed language. It also encourages many good programming practices.
- Like any other modern high-level programming language, Python is modular and supports object-oriented programming concepts. This is good for packaging and re-usage of the generated code. This often results in more transparent, maintainable and bug-free code.
- The documentation is tightly integrated with the code, and follows a Linux like help system. With the advent of integrated environments, these have taken a very good shape making Programming fun.
- Python is blessed with a large standard library, and also supplemented with a large collection of add-on packages. Most importantly, all are free of cost.

Certainly, like many other high-level languages Python too has some disadvantages. We list some of them below:

- As Python is an interpreted and dynamically typed programming language, the execution of python code can be slow. We say slow as and when we compare its execution time to the compiled statically typed programming languages, such as C and Fortran.
- Python is to a great extent decentralized. What we mean is that it is spread out, with different environment, packages and documentation at different places. This can make it harder to get started with Python. This has changed a bit, with the availability of new environments like **Anaconda**.

2.4. PYTHON FOR SCIENTIFIC COMPUTING

There are very many good reasons which makes Python ideal for scientific computing. We list a few reasons below.

- Python has a strong position in scientific computing largely because of a sizable community of users. Help is easy to come with extensive documentation on both internet and within the package.
- The prime reason for the popularity is its extensive ecosystem of scientific libraries and specific environments for each targeted group. For example:

– *numpy*: http://numpy.scipy.org – Numerical Python

– *scipy*: http://www.scipy.org – Scientific Python

– *matplotlib*: http://www.matplotlib.org – graphics library

- The performance is really awesome. This is due to the close integration with time-tested and highly optimized codes written in C and Fortran:

– blas, atlas blas, lapack, arpack, Intel MKL, . . .

- In the era of parallel processing, Python has kept pace with the developments. It has good support for

– *Parallel processing* with processes and threads;

– Interprocess communication (MPI);

– GPU computing (OpenCL and CUDA)

- Readily available and suitable for use on high-performance computing clusters.
- No license costs, no unnecessary use of research budget.

All these features make Python highly attractive for scientific computing. As a regular prac

2.4.1. Python Environments

The code in Python is interpreted. The standard implementation of the interpreter (technically referred to as CPython) is what actually runs the python code on a computer. There are also many different environments through which the python interpreter can be used. Each environment has different advantages and is suitable for different workflows. One strength of python is that it is versatile and can be used in complementary ways. In order to reduce the complexity, we introduce the core concepts in the following sections.

2.4.2. Python Interpreter

The standard way to use the Python programming language is to use the Python interpreter to run python code. The python interpreter is a program that reads and execute the python code in files passed to it as arguments. At the command prompt, the command python is used to invoke the Python interpreter. For example, to run a file hello-world.py that contains python code from the command prompt, one types:

$ python hello-world.py

We can invoke the interpreter by typing

python

at the command line. This allows you to interactively type python code into the interpreter. This is often how we want to work when developing scientific applications. It could also be the case when doing small calculations. But the standard python interpreter is not very convenient for this kind of work, due to a number of limitations. These issues have been addressed by developers at Python, which led to the launch of a versatile and a highly interactive shell for Python. We introduce IPython in the next section.

2.4.3. IPython

The need for a separate shell for Python was addressed by the launching of IPython. Basically, IPython is an interactive shell that addresses the various limitations of the standard python interpreter. Very quickly, it has become or more precisely labeled as work-horse for scientific use of python. It provides an interactive prompt to the python interpreter with a greatly improved user-friendliness. Some of the many useful features of IPython include:

- Like unix/linux there is a provision for command history, which can be browsed with the up and down arrows on the keyboard.
- Tab auto-completion. This is a very useful feature for lazy but creative programmers.
- In-line editing of code. Simply fantastic. Just try it to feel it.
- Object introspection, and automatic extract of documentation strings from python objects like classes and functions.
- Good interaction with operating system shell.
- There is also good support for multiple parallel back-end processes. These could be running on computing clusters or cloud services like Amazon EE2.

2.4.4. Jupyter (IPython) Notebook

Back to school with notebooks. This may sound a bit daunting, but yes we are heading back to school. An IPython notebook is an HTML-based notebook environment for Python. The concepts are borrowed from Mathematica or Maple. It is based on the IPython shell; but provides a cell-based environment with great interactivity, where calculations can be organized and documented in a structured way. This has been replaced by an upgraded version called Jupyter Notebook. Although using a web browser as graphical interface, IPython notebooks are usually run locally, from the same computer that run the browser. To start a new IPython notebook session, run the following command:

$ ipython notebook or

$ jupyter notebook

from a directory where you want the notebooks to be stored. This will open a new browser window (or a new tab in an existing window) with an index page where existing notebooks are shown and from which new notebooks can be created.

This is the right way to get started with notebooks. Let us start by taking baby steps into Python.

One can start by typing in Notepad (in Windows) or in vim or nano (in Linux). I would advice using Spyder IDE which comes as integral to Anaconda installation.

coding: utf-8

* All the code in Python is usually stored in text files. They are denoted with an ending/extension "`.py`":

For example,

myprogram.py

* Each line in a Python program is by default assumed to be a Python statement.

* The only exception is comment lines, which start with the character `` (optionally preceded by an arbitrary number of white-space characters, i.e., tabs or spaces) or a # (C-style). Comment lines are usually ignored by the Python interpreter.

* Python program are run from the command line by typing:

$ python myprogram.py

* It is a practice on UNIX systems to define the path to the interpreter on the first line of the program (note that this is a comment line as far as the Python interpreter is concerned):

#!/usr/bin/env python

The reader is recommended to start up an IPython session or a Jupyter notebook for executing the commands and code snippets given below.

Example:

In[1]:

ls scripts/hello-world*.py

In[2]:

cat scripts/hello-world.py

In[3]:

get_ipython().system('python scripts/hello-world.py')

The standard character encoding is ASCII. But, any other encoding can be used, for example, UTF-8. I myself prefer UTF-8. To specify that UTF-8 is used we include the special line at the top of the file.

-*- coding: UTF-8 -*-

Other than these two *optional* lines in the beginning of a Python code file, no additional code is required for initializing a program.

We take a brief tour of modules in Python in the next section.

2.4.5. Modules

Modules are core to any Python implementation. As a matter of fact, most of the functionality is embedded in *modules*. The Python Standard Library (PSL) hosts a comprehensive collection of modules. These modules provide cross-platform implementations of common facilities, which involve access to the operating system, input/output of files, management of strings, communication through networks, and many more facilities.

The way to utilize a module in Python is to first import it in the code. This is done in a standard way by using the *import* statement. Let us start with a few simple examples. Suppose you want to carry out some mathematical operations. There is a module in PSL which helps you carry out operations with much ease, as it hosts many functions in it. This module is called *math*. We can import the math functions using a simple statement as follows:

```
In[6]: import math
```

What this statement does is that it makes available the entire gamut of functions available in math module to the entire code which comes after its position. We can invoke any function available in it, by just typing the first letter, and pressing TAB button which autofills or provides all the functions available to choose from which start with that letter. So, what are you waiting for let's get started:

```
In[7]: import math
x = math.sin(math.pi)
print(x)
```

Is it convenient? Obviously not. At least in this instance. I need to prefix *math* every time I import some function or variables from *math* module. Is there a way out? Yes, there is an alternate way to import modules. We can chose to import all symbols (functions and variables) in a module, in the present (current) namespace using the keywords *from* and *with*. We show below an example, which illustrates what we mean:

```
In[8]: from math import *
a = sin(3 * pi)
print(a)
```

This may look convenient. Yes it is in this instance, where you are not coding in a big way. But, in codes which are very large this could be a problem. As the different modules may have identical function and variable names (not all of them, but there could be some which may have the same names). Importing them in the above way in a single namespace can result in a confusion as regards to its import from different modules. Hence, it is a good idea to keep the symbols from each module in their own namespaces. That means, using the `import math` is a better option for bigger programs or code. This would eliminate potentially confusing problems with *name space collisions*. I hope the reader would pay appropriate attention to detail in situations like the ones pointed out above.

As a third alternative, we can choose to import only a few selected symbols from a module by explicitly listing which ones we want to import. This is to load the namespace with only the required functions. An example below illustrates this:

```
In[9]: from math import pi, sin
z = sin(2 * pi)
```

print(z)

An obvious question which would occur is this:

How does one know all the functions listed in a given module?

After a specific module is imported, one can list all the symbols in it by using the *dir* function:

In[10]: import math

print(dir(math))

If one is interested in knowing the details of each function, one can use the *help* function. In this way we can get complete description of each function. These are known as docstrings. It is not necessary that all of the functions in a given module will have docstrings. But, this is the correct way of documenting the functions and their usage. We give a few examples below, which illustrate this:

In[11]: help(math.log)

In[12]: log(10, 1)

In[13]: log(10, 3)

We can also use the *help* function directly on modules: Try

help(math)

Below, we list some of the extremely useful modules form the Python standard library. They are:

os, sys, math, shutil, re, subprocess, multiprocessing, threading

A complete lists of standard modules for Python 2 and Python 3 are available at:

http://docs.python.org/2/library/ and *http://docs.python.org/3/library/* respectively.

2.4.6. Variables and Types

Defining variables in a code is the most obvious way of communicating with the language. Variable names should be informative and descriptive of data which they hold. In Python, variable names are allowed to contain alphanumerical characters [a-z], [A-Z], [0-9]. Also, some special characters such as [_] are allowed. In principle, normal variable names must start with a letter and not a number.

Although it is not a convention, still people resort to using lower-case letters as variable names. When we discuss Classes, it can be seen that there is a convention to use a capital letter for Class names.

We are tempted, at times, to use certain words as variable names. Python forbids using some such names. We provide a list of some keywords which should not be used as variable names. These keywords are:

and, as, assert, break, class, continue, def, del, elif, else, except,

exec, finally, for, from, global, if, import, in, is, lambda, not, or,

pass, print, raise, return, try, while, with, yield

The use of the keyword `*lambda*`, as a natural variable name is not allowed in a scientific program, being a keyword it cannot be used to declare a variable.

2.4.7. Assignment Operator

In Python the assignment is through an operator (=). Being a dynamically typed language, one need not specify the type of a variable as one is created. A variable is created by assigning a value to it:

In[14]:

```
#variable assignments
a = 10.0
my_height = 165.7
```

Every variable has a *type* associated with it, although not explicitly specified. The *type* is derived from the value that is assigned to the respective variable. You can find out the type by using *type()* function.

In[15]:

```
type(a)
```

Assigning a new value can result in a change of the type.

In[16]:

```
a = 1
```

In[17]:

```
type(a)
```

With this small introduction, we can switch gear and see how Python is useful in Bioinformatics.

For a good *Bioinformatics* background, I would suggest referring to the following two books:

Ø *Bioinformatics: Sequence and Genome Analysis*, by David W. Mount (Cold Spring Harbor Laboratory Press)

Ø *Bioinformatics and Functional Genomics*, by Jonathan Pevsner (Wiley-Blackwell)

Ø *Learning Python* – Powerful Object-Oriented Programming, by Mark Lutz (O'Reilly Media)

We begin this section with simple string manipulations, basic to any sequence analysis. It is assumed that the reader would familiarize themselves with the basics of Python. The book *Learning Python* is recommended for beginners.

2.4.8. String Operations

Let us start with the some basic operations on strings. For most part in bioinformatics, you will be dealing with strings. There are four binary operators that act on strings:

1. *in*
2. *not in*
3. +, and
4. *

The first three have both operands as that of *type* string. The last operator (*) takes an integer as the other operator. Substrings are extracted using *subscription* (one-character) or *slicing* (longer).

>>> 'TATA' in 'ATTTATATTTTTTTTAATATA'

True

>>> 'AAA' in 'ATTTATATTTTTTTTAATATA'

False

>>> 'ATTA' not in 'ATTTATATTTTTTTTAATATA'

True

One can concatenate strings to produce a new string.

>>> 'AT' + 'CG'

'ATCG'

>>> 'ACD' + 'KLY'

'ACDKLY'

You can repeat a string a certain number of times by using * operator.

```
>>> 'AT' * 6
'ATATATATATAT'
```

One can extract a one-character string by subscription.

```
>>> 'ATTTATATTTTTTTTAATATA'[0]
'A'
>>> 'ATTTATATTTTTTTTAATATA'[1]
'T'
```

One is allowed to use negative indices; the last character is at index *-1*:

```
>>> 'ATTTATATTTTTTTTAATATA'[-1]
'A'
>>> 'ATTTATATTTTTTTTAATATA'[-10]
'T'
```

You can expect *IndexError* if the index is out of range.

With slicing one can extract substrings consisting of a series of characters. Slicing is the most commonly used operation in Python and you'll be using it very often on strings.

A slice [p:q] is read as 'from character *p* up to but not including *q*.' Let us see it on a few examples:

```
>>> 'ATTTATATTTTTTTTAATATA'[1:3]
'TTT'
>>> 'ATTTATATTTTTTTTAATATA'[10:-1]
'TTTTTAATAT'
```

If both the indices are omitted, then the entire string is selected. If a third number is specified, following a colon, it indicates the number of characters to *skip* after each one that is included (known as *step*).

```
>>> 'ATTTATATTTTTTTTAATATA'[0:7:2]
'ATA'
```

There are some commonly used methods of the *str* type. Syntax of some are given below:

```
your_string.count(new_string[, start[, end]])
your_string.find(new_string[, start[, end]])
```

your_string.startswith(new_string[, start[, end]])

A few examples are given below (these are self explanatory) :

```
>>> 'ATTTATATTTTTTTTAATATA.'count('TT')
5
>>> 'ATTTATATTTTTTTTAATATA.'find('TT')
1
```

returns the first location of match.

```
>>> 'ATTTATATTTTTTTTAATATA.'find('TT', 5, 14)
7
>>> 'ATTTATATTTTTTTTAATATA.'startswith('TT')
False
>>> 'ATTTATATTTTTTTTAATATA.'startswith('TT', 1)
    True
```

Many of these built-in methods can be used for varied searches in sequences (for example, checking for recognition site etc.). Let us start with an example by *reverse transcribing a DNA* sequence. We will be writing it as a function, using the standard libraries of Python.

```
def revComp(seq, isDNA=True):
        from string import maketrans
        if isDNA:
                seq = seq.replace('U', 'T')
                translateTable = maketrans('ATGC', 'TACG')
        else:
                seq = seq.replace('T', 'U')
                translateTable = maketrans('AUGC', 'UACG')
        complement = seq.translate(translateTable)
        reverse_complement = complement[::-1]
return reverse_complement
```

This code relies on *translate* and *maketrans* functions imported from the *string* module; allowing a character substitution table to be used for equivalently positioned letters in two strings. The user-defined function *isDNA* also checks for wrong letters (U in place of T).

2.4.9. Reading Files

When reading an external file in Python, one needs to specify the *location* and *name* of the file. This is known as *specifying the path* to a file. We use python's open() function to get access to the contents of file. The open function passes a *file handle*, which is used to access and represent the file. For example,

path = 'example/sample.txt'

fo = open(path)

fo is known as file object. The function also takes an optional second argument. One can specify the *'mode'* in which the file should be opened.

Once you finish reading the file, you should close it using the close() function.

fo.close()

The file object allows to read the data in the file using functions:

read() – reads the entire file, into a single string;

readline() – one line from a file; each line ends with a newline ('\n') character;

readlines() – reads the entire lines into a list, with each line being an element in the list.

An alternative to manually closing a file in Python, is using *with* construct.

```
with open(path) as fo:
        for line in fo:
                # process line
```

2.4.10. Reading FASTA Files

FASTA is a text format to store nucleotide or protein sequences in one-letter codes. A FASTA file can have one or more sequence entries. Each entry has two parts.

a) *Header*: every sequence entry starts with a header symbol '>.' It is followed by an identifier or a comment; usually gene or protein name with database codes are provided.

b) *Sequence*: The next line is the sequence, till the next Header line is reached.

Header line is also known as comment line. An example of a FASTA sequence of a protein from the model plant *Arabidopsis thaliana* is given below:

>sp|O64587|ACD11_ARATH Accelerated cell death 11 OS=Arabidopsis thaliana GN=ACD11 PE=1 SV=1
MADSEADKPLRKISAAFKKLAIIVNSPNPEVPVTQFSHACSLVSPLF-GCLGIAFKFAEMD
YVAKVDDLVRASSSISTLVVMMDKDIEADCVRKAGSHTRNLL-RVKRGLDMVKVLFEQIIA
SEGDNSLKDPATKSYAQVFAPHHGWAIRKAVSLGMYALPTRAHLL-NMLKEDEAAAKIHMQ
SYVNSSAPLITYLDNLFLSKQLGIDW

Often one encounters many sequences, sometimes the whole genome, in a single FASTA file. We will write a function to read such a file, and generate a list of individual sequences.

```
def readFasta(name):
        fo = open(name)
        seqs = []
        seqFrags = []
        for line fo:
                if line.startswith('>'):
                        if seqFrags:
                                seq = '.'join(seqFrags)
                                seqs.append(seq)
                        seqFrags = []
                else:
                        seq = line.rstrip()
                        seqFrags.append(seq)
        if seqFrags:
                seq = '.'join(seqFrags)
                seqs.append(seq)
        fo.close()
return seqs
```

As you grow comfortable using Python and packages, you will start thinking about additional checks that need to be performed. For example, if the file is a genuine FASTA file, etc. **BioPython** is a nice package to get acquainted with for performing FASTA checks and manipulations.

2.5. REGULAR EXPRESSIONS

Regular expressions are strings. These strings are used as *patterns* to perform **match** and **search** operations on *target* strings. A successful match operation yields a *match object*, which has quite a few fields and methods providing details about the match, such as the position within the target where the pattern was found. If the match or search fails, the result is **None**, rather than a match object. Python does distinguish between *matching* and *searching*. I leave it to the inquisitive reader to delve into technicalities. We will take a simple example to illustrate the concepts involved.

Assume you are working with multiple sequences contained in a FASTA file. We will see how we can extract entries using regular expressions.

Say, you want to extract the description line or header. Then the following will do it.

```
re.search(r'^>[^>]*', src).group()
```

What it means is 'match the following > which starts a line with all characters that aren't a >.'

As an aside, this is equivalent to *grep* operation on linux command line.

We can improve our approach by capturing two groups, namely *header* and *sequence* parts.

```
re.search(r'^>(.*)$([^>]+)', src, re.M).groups()
```

I leave it to the reader to derive meaning of this regular expression.

Please refer to Appendix 1 for a collection of useful regular expressions.

The following books are suggested for the interested reader for delving into the ocean of regular expressions:

- *Mastering Regular Expressions* – Understand Your Data and Be More Productive by Jeffrey E. F. Friedl, 3rd Edition, Publisher: O'Reilly Media.
- *Regular Expression Pocket Reference* by Tony Stubblebine, Publisher: O'Reilly Media.

We have shown through many examples the utility of Python in Bioinformatics studies. The applications are many and we have covered the basics of the concepts involved. The interested student is required to acquire a good mastery in Python to write and use common and specific user-defined functions for their respective scientific investigations. Interested reader is advised to follow books for getting a deeper insight into programming in Python.

1. *Bioinformatics Programming – Using Python* by Mitchell L. Model (O'Reilly, 2010)
2. *Python Programming for Biology* by Tim J. Stevens and Wayne Boucher (Cambridge University Press, 2015).

2.6. HIGH PERFORMANCE COMPUTING

High Performance Computing (HPC) is one of the happening things in the present computational world. As a matter of fact, it is a norm in the present scientific and engineering disciplines. It is imperative to use all the available computational resources to its maximum to improve the speed at which the tasks are accomplished. We would also like to introduce High Performance Computing (HPC) concepts using IPython. Let us see how we can make use of IPython to make use of multicore architectures of normal desktops.

In[1]:

```
get_ipython().magic('matplotlib inline')
import matplotlib.pyplot as plt
```

Python has a built-in library (process-based) for concurrent computing, which is known as *multiprocessing*. We illustrate below some of the utilities of the package.

We start by importing the necessary packages.

In[2]:

```
import multiprocessing
import os
import time
import numpy
```

Let us define a function which returns process id of a given computational task.

In[3]:

```
def task(args):
    print("PID of the process = ", os.getpid(), ", args =", args)
    return os.getpid(), args
```

In[4]:

```
task("test")
```

In[5]:

```
pool = multiprocessing.Pool(processes=2)
In[6]:
result = pool.map(task, [10,20,30,40,50,60,70,80])
In[7]:
result
```

The *multiprocessing* package is an extremely useful one for tasks which need to be run in parallel, and which need not communicate with each other than when sending the initial data to the pool of processes and when and collecting the results. Let us also see if our workhorse IPython too has such a facility.

2.6.1. IPython parallel

IPython is quite versatile and interestingly it supports a parallel computing environment. Also, it is very easy to carry out parallel computing in this environment. This concept has been built upon ipython engines and controllers. The way to get started is by firing up a cluster of engines. We have to issue the *ipcluster* command, as shown below:

```
$ ipcluster start -n 2
```

You can also start this from the "Clusters" tab on the IPython notebook dashboard page. This will start 2 IPython engines on the current host. This is quite helpful on multicore systems. When installed on a computing cluster, it is possible to start IPython clusters spanning across many nodes. The reader or user is referred to the official documentation for more information:

http://ipython.org/ipython-doc/dev/parallel/

Now, let us see how to use the IPython cluster in our Python programs or notebooks. We start by creating an instance of *IPython.parallel.Client* as follows:

```
In[8]:
from IPython.parallel import Client
In[9]:
clients = Client()
```

Using the 'ids' attribute we can retrieve a list of ids for the IPython engines in the cluster:

```
In[10]:
clients.ids
```

Since, each of these engines are available to execute tasks, we can selectively run code on individual engines as follows:

```
In[11]:
def getpid():
  """ return the unique ID of the current process """
  import os
  return os.getpid()
In[12]:
# first try it on the notebook process
getpid()
In[13]:
# run it on one of the engines
clients[0].apply_sync(getpid)
In[14]:
# run it on ALL of the engines at the same time
clients[:].apply_sync(getpid)
```

So, one is using the cluster of engines to execute tasks in parallel. The best and the easiest way to dispatch a function to different engines is by defining the function with a decorator. We show an example below:

```
@view.parallel(block=True)
```

Here, *view* is supposed to be the engine pool which we want to dispatch the function (task). One uses *map* method to dispatch the function to an engine. In Python, a *decorator* is a language construct which automatically wraps the function into another function or a class. To see how all this works, lets look at an example:

```
In[15]:
direct_view = clients[:]
In[16]:
@direct_view.parallel(block=True)
def dummy_task(delay):
  """ a dummy task that takes 'delay' seconds to finish """
  import os, time
  t0 = time.time()
```

```
    pid = os.getpid()
    time.sleep(delay)
    t1 = time.time()
    return [pid, t0, t1]
```

In[17]:

```
# generate random delay times for dummy tasks
delay_times = numpy.random.rand(4)
```

Now, to map the function `dummy_task` to the random delay time data, we use the *map* method in `dummy_task`:

In[18]:

```
dummy_task.map(delay_times)
```

We now would like to illustrate the above using many more tasks and also visualize how these tasks are executed on different IPython engines:

In[19]:

```
def visualize_tasks(results):
    res = numpy.array(results)
    fig, ax = plt.subplots(figsize=(10, res.shape[1])
    yticks = []
    yticklabels = []
    tmin = min(res[:,1])
    for n, pid in enumerate(numpy.unique(res[:,0])):
        yticks.append(n)
        yticklabels.append("%d" % pid)
        for m in numpy.where(res[:,0] == pid)[0]:
            ax.add_patch(plt.Rectangle((res[m,1] – tmin, n-0.25),
                    res[m,2] – res[m,1], 0.5, color="green", alpha=0.5))

    ax.set_ylim(-.5, n+.5)
    ax.set_xlim(0, max(res[:,2]) – tmin + 0.)
    ax.set_yticks(yticks)
    ax.set_yticklabels(yticklabels)
    ax.set_ylabel("PID")
```

```
    ax.set_xlabel("seconds")
In[20]:
delay_times = numpy.random.rand(64)
In[21]:
result = dummy_task.map(delay_times)
visualize_tasks(result)
```

You see that it is a nice and easy parallelization. This way we have utilized all the engines quite well.

But one shortcoming so far is that the tasks are not load balanced, so one engine might be idle while others still have more tasks to work on.

Is there a way to handle *load balancing*? IPython parallel environment has a provision for this and it provides a number of alternative "*views*" of the engine cluster. There is a view that provides load balancing well (above we have used the "direct view", which is why we called it "direct_view").

To obtain a load balanced view we simply use the `load_balanced_view` method in the engine cluster client instance *clients*:

```
In[22]:
lbview = clients.load_balanced_view()
In[23]:
@lbview.parallel(block=True)
def dummy_task_load_balanced(delay):
    """ a dummy task that takes 'delay' seconds to finish """
    import os, time
    t0 = time.time()
    pid = os.getpid()
    time.sleep(delay)
    t1 = time.time()
    return [pid, t0, t1]
In[24]:
result = dummy_task_load_balanced.map(delay_times)
visualize_tasks(result)
```

In the example above we can see that the engine cluster is a bit more efficiently used, and the time to completion is shorter than in the previous example.

There are many other ways to use the IPython parallel environment. The official documentation has a nice guide:

http://ipython.org/ipython-doc/dev/parallel/

2.6.2. MPI

At times, some of the computations may require more communication between processes. In such cases, sophisticated solutions such as MPI and OpenMP are needed. MPI is process based parallel processing library/ protocol, and can be used in Python programs through the *mpi4py* package:

http://mpi4py.scipy.org/

To use the *mpi4py* package we issue the following statement:

from mpi4py import MPI

Typically, a MPI python program must be started using the mpirun -n N command. Here, N is the number of processes that should be included in the process group. We present a few examples below:

Example 1

In[25]:

```
get_ipython().run_cell_magic('file', 'mpitest.py', '\nfrom mpi4py import MPI\n\ncomm = MPI.COMM_WORLD\nrank = comm.Get_rank()\n\nif rank == 0:\n  data = [1.0, 2.0, 3.0, 4.0]\n  comm.send(data, dest=1, tag=11)\nelif rank == 1:\n  data = comm.recv(source=0, tag=11)\n  \nprint "rank =", rank, ", data =", data')
```

In[26]:

```
get_ipython().system('mpirun -n 2 python mpitest.py')
```

Example 2

Send a numpy array from one process to another:

In[27]:

```
get_ipython().run_cell_magic('file', 'mpi-numpy-array.py', '\nfrom mpi4py import MPI\nimport numpy\n\ncomm = MPI.COMM_WORLD\nrank = comm.Get_rank()\n\nif rank == 0:\n  data = numpy.random.rand(10)\n  comm.Send(data, dest=1, tag=13)\nelif rank == 1:\n  data = numpy.empty(10, dtype=numpy.float64)\n  comm.Recv(data, source=0, tag=13)\n  \nprint "rank =", rank, ", data =", data')
```

In[28]:

get_ipython().system('mpirun -n 2 python mpi-numpy-array.py')

Example 3: Matrix-vector multiplication

In[29]:

prepare some random data
N = 16
A = numpy.random.rand(N, N)
numpy.save("random-matrix.npy", A)
x = numpy.random.rand(N)
numpy.save("random-vector.npy", x)

In[30]:

get_ipython().run_cell_magic('file', 'mpi-matrix-vector.py', '\nfrom mpi4py import MPI\nimport numpy\n\ncomm = MPI.COMM_WORLD\nrank = comm.Get_rank()\np = comm.Get_size()\n\ndef matvec(comm, A, x):\n m = A.shape[0] / p\n y_part = numpy.dot(A[rank * m:(rank+1)*m], x)\n y = numpy.zeros_like(x)\n comm.Allgather([y_part, MPI.DOUBLE], [y, MPI.DOUBLE])\n return y\n\nA = numpy.load("random-matrix.npy")\nx = numpy.load("random-vector.npy")\ny_mpi = matvec(comm, A, x)\n\nif rank == 0:\n y = numpy.dot(A, x)\n print(y_mpi)\n print "sum(y – y_mpi) =", (y – y_mpi).sum()')

In[31]:

get_ipython().system('mpirun -n 4 python mpi-matrix-vector.py')

Example 4: Sum of the elements in a vector

In[32]:

# prepare some random data
N = 128
a = numpy.random.rand(N)
numpy.save("random-vector.npy", a)

In[33]:

get_ipython().run_cell_magic('file', 'mpi-psum.py', '\nfrom mpi4py import MPI\nimport numpy as np\n\ndef psum(a):\n r = MPI.COMM_WORLD.Get_rank()\n size = MPI.COMM_WORLD.Get_size()\n m = len(a) / size\n locsum = np.sum(a[r*m:(r+1)*m])\n rcvBuf = np.array(0.0, \'d\')\n MPI.COMM_WORLD.Allreduce([locsum, MPI.DOUBLE], [rcvBuf, MPI.DOUBLE], op=MPI.SUM)\n return rcvBuf\n\na = np.load("random-vector.npy")\ns = psum(a)\n\nif MPI.COMM_WORLD.Get_rank() == 0:\n print "sum =", s, ", numpy sum =", a.sum()')

In[34]:

get_ipython().system('mpirun -n 4 python mpi-psum.py')

Further reading
http://mpi4py.scipy.org
http://mpi4py.scipy.org/docs/usrman/tutorial.html
https://computing.llnl.gov/tutorials/mpi/

2.7. CONCLUSIONS

In this chapter we have introduced Python in a very short and concise way. This has been done by means of examples. An interactive approach to understand the core concepts in Python are presented using IPython. At the end, a small section is devoted to describing the multiprocessing in Python, in the context of High Performance Computing.

3
CHAPTER

SEQUENCE ALIGNMENTS IN BIOLOGY

CONTENTS

We introduce some of the most common and regularly used concepts in biology and bioinformatics. This involves comparing two protein or nucleic acid sequences. We start with Local Alignments, introducing BLAST.

The identification of sequences in biology is one of the primary focuses in Bioinformatics. Several tools are available for this, each with their own algorithms and approaches. Programs like BLAST, FASTA, HMMER, and many more are being released as we are writing this chapter. The basic approach involves searching a database of curated sequences for a sequence of your interest for potential matches. The size of sequence databases is growing exponentially, with the dramatic drop in the prices of sequencing technologies. The sequencing results are helpful and also add to understanding the complex biology of living organisms. Manual interpretation is out of question. Also, the search tools come with their own statistics, conventions, and output formats. We introduce in this chapter the basic sequence search tool. A good command on BLAST will also help in comprehending the results obtained by many other sequence search tools, apart from BLAST results.

3.1. LOCAL ALIGNMENT

A *local alignment* between *s* and *t* is an alignment between a substring of *s* and a substring of *t*.

The above given definition is the simplest possible one. We will now venture into understanding this at a deeper level.

In this section we present an algorithm to find the highest scoring local alignments between two sequences. This algorithm is a variation of the basic algorithm, discussed in the previous section. The main data structure is, as before, an $(m + 1)$ x $(n + 1)$ array. Only this time the interpretation of the array values is different. Each entry (i, j) will hold the highest score of an alignment between a *suffix* of $s[1..i]$ and a *suffix* of $t[1..j]$. The first row and the first column are initialized with zeros.

For any entry (i, j), there is always the alignment between the empty suffixes of $s[1..i]$ and $t[1..j]$, which has score zero; therefore the array will have all entries greater than or equal to zero. This explains in part the initialization above.

Following initialization, the array can be filled in the usual way, with $a[i, j]$ depending on the value of three previously computed entries. The resulting recurrence is

$a[i, j] = \max$

that is, the same as in the basic algorithm, except that now we have a fourth possibility, not available in the global case, of an empty alignment.

In the end, it suffices to find the maximum entry in the whole array. This will be the score of an optimal local alignment. Any entry containing this value can be used as a starting point to get such an alignment. The rest of the alignment is obtained tracing back as usual, but stopping as soon as we reach an entry with no arrow going out. Alternatively, we can stop as soon as we reach an entry with value zero.

In general, when doing local comparison, we are interested not only in the optimal alignments, but also in near optimal alignments with scores above a certain threshold. References to methods for retrieving near optimal alignments are given at the end of this chapter.

Let us now take up the popular local alignment tool BLAST.

3.2. BLAST

BLAST is an acronym for Basic Local Alignment Search Tool. It is important to observe that here the term *database* refers simply to a usually large set of catalogued sequences. It does not imply any extra capabilities of fast access, data sharing, and so on, commonly found in standard database management systems. For us, therefore, this "database" is merely a collection of sequences, although sequence information is copiously complemented with additional information such as the origin of the data, bibliographic references, sequence function (if known), and others. BLAST returns a list of *high-scoring segment pairs* between the query sequence and sequences in the database. Before explaining what this means and how BLAST obtains its results, we will briefly introduce some terminology on segment pairs to keep this discussion consistent with the original paper describing BLAST.

A *segment* is a substring of a sequence. Given two sequences, a *segment pair* between them is a pair of segments of the same length, one from each sequence. Because the substrings in a segment pair have the same length, we can form a gapless alignment with them. This alignment can be scored using a matrix of substitution scores. No gap-penalty functions are needed, as there are no gaps. The score thus obtained is by definition the score of the segment pair. An example of a segment pair scored with the PAM120 substitution matrix is given below

K	A	L	M	R
V	A	K	N	S
-4	3	-4	-3	-1

This gives a total of -9.

We are now in possession of all the information necessary to describe precisely what BLAST does. Let us divide the process into the following steps for our comprehension:

- Given a query sequence, BLAST returns all segment pairs between the query and a database sequence with scores above a certain threshold *S*.
- The parameter *S* can be set by the user, although a default value is provided in most servers that run the program.
- We have mentioned that a characteristic of reported alignments is the absence of gaps. This is actually a major reason why BLAST is so fast, since looking for good alignments with gaps is a good deal more time consuming.
- A *maximum segment pair* (MSP) between two sequences is a segment pair of maximum score. This score is a measure of sequence similarity and can be computed precisely by dynamic programming. However, BLAST estimates this number much faster than any dynamic programming method.
- The program also returns *locally* maximal segment pairs, that is, those that cannot be improved further by extending or shortening them.

3.3. WORKING OF BLAST ALGORITHM

The BLAST approach to computing high-scoring segment pairs is as follows. It finds certain "seeds," which are very short segment pairs between the query and a database sequence. These seeds are then extended in both directions, without including gaps, until the maximum possible score for extensions of this particular seed is reached. Not all extensions are looked at. The program has a criterion to stop extensions when the score falls below a carefully computed limit. There is a very small chance of the right extension not being found due to this time optimization, but in practice this trade off is highly acceptable.

We can now narrow the steps of BLAST to just three as given below (in simple English):

- Compile a list of high-scoring strings (or words in BLAST jargon)
- Search for hits – each hit gives a seed.
- Extend seeds.

We explain this for protein sequences and then take an example for illustration. For protein sequences, the list of high-scoring words consists of all words with *w* characters (called tu-mers) that score at least *T* with some *w*-mer of the query, using a PAM matrix to compute scores. Here *w* and *T* are program parameters. Note that this list may not contain all query *w*-mers! If a query *w*-mer consists of very common amino acids, it may be left out because even its score with itself may fall below *T*. However, there is an option to force inclusion of all query w-mers. The recommended value of *w,* the seed size, is 4 for protein searches.

Two approaches are tried for scanning the database in search for hits in the list constructed in the previous step. One of them is to arrange the list words into a hash table. Then, for each database word of size *w,* it is easy to get their index in the table and compare it to the words there, which will be a small fraction of all list words.

The second method uses a *deterministic finite automaton* to search for hits. This device has states and transitions and operates like a machine. It begins in a fixed initial state, and for each character in the database a transition is made to another state. Depending on the state and on the transition, a word from the list is recognized. The automaton is built only once, using the list of high-scoring words as input, and is a compact way of storing all these words. The search is fast, as it requires only a transition per character.

The final extension is straightforward. As we mentioned, to save time the algorithm stops when the score falls a certain distance below the best one obtained to that point for shortest extensions. This is done in both directions, and the high-scoring segment pair originated from this seed is kept. There is a small probability of missing important extensions, but it is negligible.

Extension is based on a well-founded statistical theory that gives exact distribution of gapless local maximum score for random sequences, and permits a very accurate computation of the probability that the segment pair found could be possible due to chance alone. The smaller this probability, the more significant is the match.

We briefly sketch now the main points of this statistical theory. The distribution of the MSP score (defined above) for random sequences *s* and *t* of lengths *m* and *n,* respectively, can be accurately approximated as described next. This approximation gets better as *m* and *n* increase.

With this basic theory in mind we will now take up illustration. Let us delve into some practical situations, wherein you will be using the concepts learnt so far. Actually, we would take an approach where we will do this by examples.

3.4. ACCESSING DATABASES

We will look at how to access online repositories from your computers. We use **BioPython** for this. You can also look at packages available under **Bioconductor** in **R**, for accomplishing the same.

from Bio import Entrez, Medline, SeqIO

Entrez.email = "your_mail_address@here"

```
#This gives you the list of available databases
handle = Entrez.einfo()
rec = Entrez.read(handle)
print(rec)
{'DbList': ['pubmed', 'protein', 'nuccore', 'nucleotide', 'nucgss', 'nucest', 'structure', 'sparcle', 'genome', 'annotinfo', 'assembly', 'bioproject', 'biosample', 'blastdbinfo', 'books', 'cdd', 'clinvar', 'clone', 'gap', 'gapplus', 'grasp', 'dbvar', 'gene', 'gds', 'geoprofiles', 'homologene', 'medgen', 'mesh', 'ncbisearch', 'nlmcatalog', 'omim', 'orgtrack', 'pmc', 'popset', 'probe', 'proteinclusters', 'pcassay', 'biosystems', 'pccompound', 'pcsubstance', 'pubmedhealth', 'seqannot', 'snp', 'sra', 'taxonomy', 'biocollections', 'unigene', 'gencoll', 'gtr']}
from Bio.Blast import NCBIWWW
result_handle = NCBIWWW.qblast("blastn", "nt", "8332116")
```

Now that we've got a handle, we are ready to parse the output. The code to parse it is really quite small. If you expect a single BLAST result (i.e., you used a single query):

```
from Bio.Blast import NCBIXML
blast_record = NCBIXML.read(result_handle)
Please refer to BioPython documentation for parsing BLAST output.
```

Let us take another illustration. We are in search of gene coding for *insulin* in homo sapiens. This can be done by specifying the Gene and organism name in the search call to the remote online repository.

```
handle = Entrez.esearch(db="nucleotide", term='INS[Gene Name] AND "Homo sapiens"[Organism]')
rec_list = Entrez.read(handle)
if rec_list['RetMax'] < rec_list['Count']:
handle = Entrez.esearch(db="nucleotide", term='INS[Gene Name] AND "Homo sapiens"[Organism]', retmax=rec_list['Count'])
rec_list = Entrez.read(handle)
id_list = rec_list['IdList']
hdl = Entrez.efetch(db='nucleotide', id=id_list, rettype='gb')
# parsing each record found in the above search
recs = list(SeqIO.parse(hdl, 'gb'))
# print the records (optional); still a good practice to check the results
recs
[SeqRecord(seq=Seq('AGCCCTCCAGGACAGGCTGCATCAGAAGAGGCCATCAAGCAGGTCTGTTCCAAG...AAA', IUPACAmbiguousDNA()), id='NM_001291897.1', name='NM_001291897', description='Homo sapiens insulin (INS), transcript variant 4, mRNA.', dbxrefs=[]), SeqRecord(seq=Seq('AGCCCTCCAGGACAGGCTGCATCAGAAGAGGCCATCAAGCAGGTCTGTTCCAAG...AAA', IUPACAmbiguousDNA()), id='NM_001185098.1', name='NM_001185098', description='Homo sapiens insulin (INS), transcript variant 3, mRNA.', dbxrefs=[]), SeqRecord(seq=Seq('AGCCCTCCAGGACAGGCTGCATCAGAAGAGGCCATCAAGCAGGTCTGTTCCAAG...AAA', IUPACAmbiguousDNA()), id='NM_001185097.1', name='NM_001185097',
```

description='Homo sapiens insulin (INS), transcript
variant 2, mRNA.', dbxrefs=[]), SeqRecord(seq=Seq('A
GCCCTCCAGGACAGGCTGCATCAGAAGAGGCCATCAAGCAGATCACTGTC
CTT...AAA', IUPACAmbiguousDNA()), id='NM_000207.2',
name='NM_000207', description='Homo sapiens insulin
(INS), transcript variant 1, mRNA.', dbxrefs=[]),
SeqRecord(seq=Seq('GAGGTGCGGATCCTGGGCGGCCAGGGAA
GGTCTCTGCCGCCAGGGAAGTGTCCC...CTC', IUPACAmbigu-
ousDNA()), id='NG_050578.1', name='NG_050578',
description='Homo sapiens INS-IGF2 readthrough (INS-
IGF2), RefSeqGene on chromosome 11.', dbxrefs=[]),
SeqRecord(seq=Seq('GGCGGCCAGGGAAGGTCTCTGCCGCCAG
GGAAGTGTCCCAGAGACCCCTGGAGG...GCC', IUPACAmbigu-
ousDNA()), id='NG_007114.1', name='NG_007114',
description='Homo sapiens insulin (INS), RefSeq-
Gene on chromosome 11.', dbxrefs=[]), SeqRecord-
(seq=Seq('CTGGGGACCTGACCCAGCCGCAGCCTTTGTGAACCAA
CACCTGTGCGGCTCACA...CTA', IUPACAmbiguousDNA()),
id='JF909299.1', name='JF909299', description='Homo
sapiens insulin (INS) mRNA, partial cds.',
dbxrefs=[]), SeqRecord(seq=Seq('GTACAAAAAGCAGGCTC
CACCATGGCCCTGTGGATGCGCCTCCTGCCCCTGCT...TAC', IUPA-
CAmbiguousDNA()), id='DQ896283.2', name='DQ896283',
description='Synthetic construct Homo sapiens clone
IMAGE:100010743; FLH192918.01L; RZPDo839A1068D in-
sulin (INS) gene, encodes complete protein.',
dbxrefs=[]), SeqRecord(seq=Seq('CTCGAGGGGCCTAGACAT
TGCCCTCCAGAGAGAGCACCCAACACCCTCCAGGCT...TCA', IUPA-
CAmbiguousDNA()), id='AH002844.2', name='AH002844',
description='Homo sapiens insulin (INS) gene, com-
plete cds.', dbxrefs=[]), SeqRecord(seq=Seq('GAGC
AGGGCTGGACCTGTGAGCCCAGGTCACAGATGAGAAAACCGACCCCTG-
GT...AAG', IUPACAmbiguousDNA()), id='AH012037.2',
name='AH012037', description='Homo sapiens haplo-
type ICa tyrosine hydroxylase (TH) gene, partial
sequence; insulin (INS) gene, complete cds; and
insulin-like growth factor 2 (IGF2) gene, partial
sequence.', dbxrefs=[]), SeqRecord(seq=UnknownS
eq(135086622, alphabet = IUPACAmbiguousDNA(), char-

acter = 'N'), id='NC_000011.10', name='NC_000011', description='Homo sapiens chromosome 11, GRCh38.p7 Primary Assembly.', dbxrefs=['BioProject:PRJNA168', 'Assembly:GCF_000001405.33']), SeqRecord(seq= UnknownSeq(134889443, alphabet = IUPACAmbiguousDNA(), character = 'N'), id='NC_018922.2', name='NC_018922', description='Homo sapiens chromosome 11, alternate assembly CHM1_1.1, whole genome shotgun sequence.', dbxrefs=['BioProj ect:PRJNA178030', 'BioSample:SAMN02205338', 'Assembly:GCF_000306695.2']), SeqRecord(seq=Seq('G TTCGTTGCAACAAATTGATGAGCAATGCTTTTTATAATGCC AACTTTG-TACA...AAC', IUPACAmbiguousDNA()), id='KR710185.1', name='KR710185', description='Synthetic construct Homo sapiens clone CCSBHm_00010262 INS (INS) mRNA, encodes complete protein.', dbxrefs=[]), SeqRecord(seq=Seq ('GTTCGTTGCAACAAATTGATGAGCAAT-GCTTTTTATAATGCCAACTTTGTACA...AAC', IUPACAmbiguousDNA()), id='KR710184.1', name='KR710184', description='Synthetic construct Homo sapiens clone CCSBHm_00010257 INS (INS) mRNA, encodes complete protein.', dbxrefs=[]), SeqRecord (seq=Seq('GTTC GTTGCAACAAATTGATGAGCAATGCTTTTTATAATGCCAACTTTGTA-CA...AAC', IUPACAmbiguousDNA()), id='KR710183.1', name='KR710183', description='Synthetic construct Homo sapiens clone CCSBHm_00010256 INS (INS) mRNA, encodes complete protein.', dbxrefs=[]), SeqRecord(seq=Seq ('GTTCGTTGCAACAAATTGATGAGCAAT-GCTTTTTATAATGCCAACTTTGT ACA...AAC', IUPACAmbiguousDNA()), id='KR710182.1', name='KR710182', description='Synthetic construct Homo sapiens clone CCSBHm_00010255 INS (INS) mRNA, encodes complete protein.', dbxrefs=[]), SeqRecord(seq=Unkn ownSeq(132164248, alphabet = IUPACAmbiguousDNA(), character = 'N'), id='CM000262.1', name='CM000262', description='Homo sapiens chromosome 11, whole genome shotgun sequence.', dbxrefs=['BioProject:PRJNA1431', 'BioSample:SAMN02981219']), SeqRecord(seq=Unknown

Seq(3852046, alphabet = IUPACAmbiguousDNA(), character = 'N'), id='CH471158.1', name='CH471158', description='Homo sapiens 211000035842702 genomic scaffold, whole genome shotgun sequence.', dbxrefs=[' BioProject:PRJNA1431', 'BioSample:SAMN02981219']), SeqRecord(seq=Seq ('AGCCCTCCAGGACAGGCTGCATCAGAAGAGGCCATCAAGCAGATCACTGTCCTT...AAA', IUPACAmbiguousDNA()), id='BC005255.1', name='BC005255', description='Homo sapiens insulin, mRNA (cDNA clone MGC:12292 IMAGE:3950204), complete cds.', dbxrefs=[]), SeqRecord(seq=UnknownSeq(606048, alphabet = IUPACAmbiguousDNA(), character = 'N'), id='NT_086367.4', name='NT_086367', description='Homo sapiens chromosome 11 sequence, ENCODE region ENm011.', dbxrefs=[]), SeqRecord(seq=Seq ('GCTGCATCAGAAGAGGCCATCAAGCACATCACTGTCCTTCTGCCATGGCCCTGT...ATG', IUPACAmbiguousDNA()), id='JQ951950.1', name='JQ951950', description='Homo sapiens preproinsulin (INS) mRNA, complete cds.', dbxrefs=[]), SeqRecord (seq=Seq(' CTGCCAGCACTGCCTGTCTGAGGAGCCTGAGATTCGGGCCTTCGACCCTG AGGC...TCA', IUPACAmbiguousDNA()), id='L15440.1', name='HUMINSTHIG', description="Homo sapiens tyrosine hydroxylase (TH) gene, 3' end; insulin (INS) gene, complete cds; insulin-like growth factor 2 (IGF2) gene, 5' end.", dbxrefs=[]), SeqRecord(seq=Seq ('GCGATCGCCATGGCCCTGTGGATGCGCCTCCTG CCCCTGCTGGCGCTGCTGGCC...AAC', IUPACAmbiguousDNA()), id='AB587580.1', name='AB587580', description='Synthetic construct DNA, clone: pF1KB8864, Homo sapiens INS gene for insulin, without stop codon, in Flexi system.', dbxrefs=[]), SeqRecord(seq=Seq('CTGGGGCTGCTGTCCTAAGGCAGGGTGGAACT AGGCAGCCAGCAGGGAGGGGA...CCC', IUPACAmbiguousDNA()), id='M10039.1', name='HUMINSPR', description="Human alpha-type insulin gene and 5' flanking polymorphic region.", dbxrefs=[]), SeqRecord(seq=Seq('TA GGGGCTCAGCCCAGATGACACTATGGGGGTGATGGTGTCATGGGACCTGGCC...CTC', IUPACAmbiguousDNA()), id='J00267.1', name='HUMINS1UB', description="Human insulin

```
gene: beta allele 5'flank (ullrich).", dbxrefs=[]),
SeqRecord(seq=Seq('TAGGGGCTCA        GCCCAGATGACAC-
TATG GGGGTGATGGTGTCATGGGACCTGGCC...CTC', IUPACAm-
biguousDNA()),  id='J00266.1',  name='HUMINS1UA',
description="Human insulin gene: alpha allele 5'flank
(ullrich).", dbxrefs=[]), SeqRecord(seq=Seq('GTAC
AAAAAAGCAGGCTCCACCATGGCCCTGTGGATGCGCCTCCTGCCCCTG
CT...TAC', IUPACAmbiguousDNA()), id='DQ893040.2',
name='DQ893040',    description='Synthetic    con-
struct clone IMAGE:100005670; FLH192922.01X; RZ-
PDo839A1078D insulin (INS) gene, encodes complete
protein.', dbxrefs=[]), SeqRecord(seq=Seq('AGCAG
GTCTGTTCCAAGGGCCTTTGCGTCAGGTGGGCTCAGGGTTCCAGGGT
GG...AGC', IUPACAmbiguousDNA()), id='AJ009655.1',
name='AJ009655',  description='Homo  sapiens  ins
gene, partial.', dbxrefs=[])]

# you can pull out a record of interest from the
above list; typically we do this if we are inter-
ested in a specific gene identifier
for rec in recs:
if rec.name == 'NM_000207':
     break
print(rec.name)              # print record name
print(rec.description)    # print description of
the gene
NM_000207 Homo sapiens insulin (INS), transcript
variant 1, mRNA.
# print different features of the gene(s)
for feature in rec.features:
     if feature.type == 'gene':
          print(feature.qualifiers['gene'])
     elif feature.type == 'exon':
          loc = feature.location
          print('Exon', loc.start, loc.end, loc.
strand)
     else:
```

```
        print('not processed:\n%s' % feature)
```

not processed: type: source location: [0:469] (+) qualifiers: Key: chromosome, Value: ['11'] Key: db_xref, Value: ['taxon:9606'] Key: map, Value: ['11p15.5'] Key: mol_type, Value: ['mRNA'] Key: organism, Value: ['Homo sapiens'] ['INS'] Exon 0 42 1 not processed: type: STS location: [15:186](+) qualifiers: Key: db_xref, Value: ['UniSTS:155248'] Key: gene, Value: ['INS'] Key: gene_synonym, Value: ['IDDM; IDDM1; IDDM2; ILPR; IRDN; MODY10'] Key: standard_name, Value: ['GDB:181496'] Exon 42 246 1 not processed: type: CDS location: [59:392](+) qualifiers: Key: codon_start, Value: ['1'] Key: db_xref, Value: ['CCDS:CCDS7729.1', 'GeneID:3630', 'HGNC:HGNC:6081', 'MIM:176730'] Key: gene, Value: ['INS'] Key: gene_synonym, Value: ['IDDM; IDDM1; IDDM2; ILPR; IRDN; MODY10'] Key: note, Value: ['proinsulin; preproinsulin'] Key: product, Value: ['insulin preproprotein'] Key: protein_id, Value: ['NP_000198.1'] Key: translation, Value: ['MALWMRLLPLLALLALWGPDPAAAFVNQHLCGSHLVEALYLVCGERGFFYTPKTRREAEDLQVGQVELGGGPGAGSLQPLALEGSLQKRGIVEQCCTSICSLYQLENYCN'] not processed: type: sig_peptide location: [59:131](+) qualifiers: Key: gene, Value: ['INS'] Key: gene_synonym, Value: ['IDDM; IDDM1; IDDM2; ILPR; IRDN; MODY10'] Key: inference, Value: ['COORDINATES: ab initio prediction:SignalP:4.0'] not processed: type: proprotein location: [131:389](+) qualifiers: Key: gene, Value: ['INS'] Key: gene_synonym, Value: ['IDDM; IDDM1; IDDM2; ILPR; IRDN; MODY10'] . Key: product, Value: ['proinsulin'] not processed: type: mat_peptide location: [131:221](+) qualifiers: Key: gene, Value: ['INS'] Key: gene_synonym, Value: ['IDDM; IDDM1; IDDM2; ILPR; IRDN; MODY10'] Key: product, Value: ['insulin B chain'] not processed: type: mat_peptide location: [227:320](+) qualifiers: Key: gene, Valueue: ['INS'] Key: gene_synonym, Value: ['IDDM; IDDM1; IDDM2; ILPR; IRDN; MODY10'] Key: product, Value: ['C-peptide'] not processed: type: mat_peptide location: [326:389](+) qualifiers: Key: gene, Value: ['INS'] Key: gene_synonym, Value: ['IDDM; IDDM1; IDDM2; ILPR; IRDN; MODY10'] Key: product, Value: ['insulin A chain'] not processed: type: STS location: [59:391](+) qualifiers: Key: db_xref, Value: ['UniSTS:270424'] Key: gene, Value: ['INS'] Key: gene_synonym, Value: ['IDDM; IDDM1; IDDM2; ILPR; IRDN; MODY10'] Key: standard_name, Value: ['PMC123023P3'] not processed: type: STS location: [67:368](+) qualifiers: Key: db_xref, Value: ['UniSTS:272021'] Key: gene, Value: ['INS'] Key: gene_synonym, Value: ['IDDM; IDDM1; IDDM2; ILPR; IRDN; MODY10'] Key: standard_name, Value: ['PMC21231P1'] not processed: type: STS location: [71:371](+) qualifiers: Key: db_xref, Value: ['UniSTS:259556'] Key: gene, Value: ['INS'] Key: gene_synonym, Value: ['IDDM; IDDM1; IDDM2; ILPR; IRDN; MODY10'] Key: standard_name, Value: ['Ins2'] not processed: type: STS location: [78:390](+) qualifiers: Key: db_xref, Value: ['UniSTS:256685'] Key: gene, Value: ['INS'] Key: gene_synonym, Value: ['IDDM; IDDM1; IDDM2; ILPR; IRDN; MODY10'] Key: standard_name, Value: ['Ins1'] not processed: type: STS location: [197:385](+) qualifiers: Key: db_xref, Value: ['UniSTS:265494'] Key: gene, Value: ['INS'] Key: gene_synonym, Value: ['IDDM; IDDM1; IDDM2; ILPR; IRDN; MODY10'] Key: standard_name, Value: ['PMC24644P6'] Exon 246 465 1 not processed: type: STS location: [246:465](+) qualifiers: Key: db_xref, Value: ['UniSTS:155046'] Key: gene, Value: ['INS'] Key: gene_synonym, Value: ['IDDM; IDDM1; IDDM2; ILPR; IRDN; MODY10'] Key: standard_name, Value: ['GDB:179433'] not processed: type: regulatory location: [445:451](+) qualifiers: Key: gene, Value: ['INS'] Key: gene_synonym, Value:

```
['IDDM; IDDM1; IDDM2; ILPR; IRDN; MODY10']     Key:
regulatory_class, Value: ['polyA_signal_sequence']
not processed: type: polyA_site location: [464:465]
(+) qualifiers:    Key: gene, Value: ['INS']    Key:
gene_synonym, Value: ['IDDM; IDDM1; IDDM2; ILPR;
IRDN; MODY10']
```

```
for name, value in rec.annotations.items():
    print('%s=%s' % (name, value))
```

comment=REVIEWED REFSEQ: This record has been curated by NCBI staff. The reference sequence was derived from BC005255.1 and BM510748.1. This sequence is a reference standard in the RefSeqGene project. On Jun 15, 2006 this sequence version replaced gi:4557670.

Summary: After removal of the precursor signal peptide, proinsulin is post-translationally cleaved into three peptides: the B chain and A chain peptides, which are covalently linked via two disulfide bonds to form insulin, and C-peptide. Binding of insulin to the insulin receptor (INSR) stimulates glucose uptake. A multitude of mutant alleles with phenotypic effects have been identified. There is a read-through gene, INS-IGF2, which overlaps with this gene at the 5' region and with the IGF2 gene at the 3' region. Alternative splicing results in multiple transcript variants. [provided by RefSeq, Jun 2010]. Transcript Variant: This variant (1) represents the shortest variant. All variants encode the same protein. Publication Note: This RefSeq record includes a subset of the publications that are available for this gene. Please see the Gene record to access additional publications. SAMEA2153347 [ECO:0000348] COMPLETENESS: complete on the 3' end. source=Homo sapiens (human) taxonomy=['Eukaryota', 'Metazoa', 'Chordata', 'Craniata', 'Vertebrata', 'Euteleostomi', 'Mammalia', 'Eutheria', 'Euarchontoglires', 'Primates', 'Haplorrhini', 'Catarrhini', 'Hominidae', 'Homo'] structured_ comment=defaultdict(<type 'dict'>, {'Evidence-Data': {'Transcript exon combination': 'BC005255.1, BU075674.1 [ECO:0000332]', 'RNAseq introns': 'single sample supports all introns'}}) keywords=['RefSeq'] refer ences=[Reference(title='Elaboration of a finite element model of pancreatic islet dielectric response to gap junction expression and insulin release', ...), Reference(title='Upregulation of INS-IGF2 read-through expression and identification of a novel INS-IGF2 splice variant in insulinomas', ...), Reference(title='The influence of INS VNTR class III allele on auxological parameters, glucose, insulin, lipids, and adipocytokines secretion in prepubertal children born small for gestational age', ...),

Reference(title="Allele-specific recognition of the 3' splice site of INS intron 1", ...), Reference(title='Global haplotype diversity in the human insulin gene region', ...), Reference(title='Permanent Neonatal Diabetes Mellitus', ...), Reference(title='Nucleotide sequence of a cDNA clone encoding human preproinsulin', ...), Reference(title='A structurally abnormal insulin causing human diabetes', ...), Reference(title='Effects of altered thyroid function on galactosyl diacylglycerol metabolism in myelinating rat brain', ...), Reference(title='Chemistry of insulin; determination of the structure of insulin opens the way to greater understanding of life processes', ...)] accessions=['NM_000207'] data_file_division=PRI date=30-APR-2017 organism=Homo sapiens sequence_version=2

3.5. RUNNING BLAST OVER THE INTERNET

We use the function qblast() in the **Bio.Blast.NCBIWWW** module to call the online version of BLAST. This has three non-optional arguments:

- The first argument is the blast program to use for the search, as a lower case string. The options and descriptions of the programs are available at http://www.ncbi.nlm.nih.gov/BLAST/blast_program.shtml. Currently qblast only works with blastn, blastp, blastx, tblast and tblastx.
- The second argument specifies the databases to search against. Again, the options for this are available on the NCBI web pages at http://www.ncbi.nlm.nih.gov/BLAST/blast_databases.shtml.
- The third argument is a string containing your query sequence. This can either be the sequence itself, the sequence in fasta format, or an identifier like a GI number.

The qblast function also take a number of other option arguments which are basically analogous to the different parameters you can set on the BLAST web page. We'll just highlight a few of them here:

- The qblast function can return the BLAST results in various formats, which you can choose with the optional *format_type* keyword: "HTML", "Text", "ASN.1", or "XML." The default is "XML", as that is the format expected by the parser; and
- The argument *expect* sets the expectation or e-value threshold.

For more about the optional BLAST arguments, we refer you to the NCBI's own documentation, or that built into Biopython:

```
In [1]:
from Bio.Blast import NCBIWWW
```

```
help(NCBIWWW.qblast)
Help on function qblast in module Bio.Blast.NCBI-
WWW:
qblast(program, database, sequence, auto_
format=None, composition_based_statistics=None,
db_genetic_code=None, endpoints=None, en-
trez_query='(none)', expect=10.0, filter=None,
gapcosts=None, genetic_code=None, hitlist_size=50,
i_thresh=None, layout=None, lcase_mask=None,
matrix_name=None, nucl_penalty=None, nucl_
reward=None, other_advanced=None, perc_ident=None,
phi_pattern=None, query_file=None, query_believe_
defline=None, query_from=None, query_to=None,
searchsp_eff=None, service=None, threshold=None,
ungapped_alignment=None, word_size=None, align-
ments=500, alignment_view=None, descriptions=500,
entrez_links_new_window=None, expect_low=None, ex-
pect_high=None, format_entrez_query=None, format_
object=None, format_type='XML', ncbi_gi=None, re-
sults_file=None, show_overview=None, megablast=None)
Do a BLAST search using the QBLAST server at NCBI.
Supports all parameters of the qblast API for Put
and Get.
Some useful parameters:

- program        blastn, blastp, blastx, tblastn,
or tblastx (lower case)     - database      Which
database to search against (e.g., "nr").
- sequence       The sequence to search.
- ncbi_gi        TRUE/FALSE whether to give 'gi'
identifier.
- descriptions   Number of descriptions to show.
Def 500.
- alignments     Number of alignments to show. Def
500.
- expect         An expect value cutoff. Def 10.0.
- matrix_name    Specify an alt. matrix (PAM30,
PAM70, BLOSUM80, BLOSUM45).
- filter         "none" turns off filtering. Default
no filtering
```

```
- format_type    "HTML", "Text", "ASN.1", or "XML."
Def. "XML."
- entrez_query   Entrez query to limit Blast search
- hitlist_size   Number of hits to return. Default
50
- megablast        TRUE/FALSE whether to use MEga
BLAST algorithm (blastn only)
- service         plain, psi, phi, rpsblast, mega-
blast (lower case)          This function does no
checking of the validity of the parameters    and
passes the values to the server as is.  More help
is available at:      http://www.ncbi.nlm.nih.gov/
BLAST/Doc/urlapi.html
```

Please note that the default settings on the NCBI BLAST website are not the same as the defaults on QBLAST. Please check the parameters in both the cases to get identical results (e.g., the expectation value threshold and the gap values).

We given an example of a BLAST with nucleotide sequence against nucleotide database (nt) using BLASTN. You need to know the GI number of your query sequence for carrying out this BLAST session. An example is given below:

```
In [2]:
from Bio.Blast import NCBIWWW
result_handle  =  NCBIWWW.qblast("blastn",  "nt",
"8332116")
```

Another approach is to use a query sequence, provided we are having the same is FASTA format. We need to open the file and read the record as a string and use that as the query argument:

```
In [ ]:
from Bio.Blast import NCBIWWW
fasta_string  =  open("data/some_sequence.fasta").
read()
result_handle = NCBIWWW.qblast("blastn", "nt", fas-
ta_string)
```

Another alternative to the above two approaches, is to have read the FASTA file as a SeqRecord and then supplied just the sequence itself:

```
In [ ]:
from Bio.Blast import NCBIWWW
from Bio import SeqIO
record = SeqIO.read("data/m_cold.fasta", format="fasta")
result_handle = NCBIWWW.qblast("blastn", "nt", record.seq)
```

BLAST assigns an identifier automatically, when you do this. You might prefer to use the SeqRecord object's *format* method to make a FASTA string (which will include the existing identifier):

```
In [ ]:
from Bio.Blast import NCBIWWW
from Bio import SeqIO
record = SeqIO.read("data/m_cold.fasta", format="fasta")
result_handle = NCBIWWW.qblast("blastn", "nt", record.format("fasta"))
```

The results will be available in result_handle. The typical output will be in XML format.

As a next step, you need to parse the XML output. You can do it using any language of your choice. Since, we are using Python we can attempt it using Python itself. It is also a good idea to save a local copy of the output file. This not only saves valuable time, but also makes the result usable in many other computational pipelines. Let us take a quick look at how to save the output.

3.6. SAVING BLAST OUTPUT

There is a limitation with regard to calling the result_handle multiple times in BioPython. It can be called only once, the next time it will return an empty string. With this in mind, let us save the output to a file which can be invoked at any point in time for analysis.

```
In [ ]:
with open("my_blast-result.xml", "w") as save_to:
     save_to.write(result_handle.read())
```

```
    result_handle.close()
```

The results are stored in the file my_blast-result.xml. Since all the data was extracted from the result_handle it was closed. The *parse* function of the BLAST parser can open a file, by using a file-handle object, hence we can just open the saved file for input:

```
In [ ]:
with open("my_blast-result.xml") as result_handle:
    print(result_handle)
```

Now that we've got the BLAST results back into a handle again, we are ready to do something with them, so this leads us right into the parsing section. You may want to jump ahead to that now ….

3.7. GRAPHICS INCLUDING GENOMEDIAGRAM

```
In [1]:
#Lets load notebook's Image
from IPython.core.display import Image
In [2]:
from reportlab.lib import colors
from reportlab.lib.units import cm
from Bio.Graphics import GenomeDiagram
from Bio import SeqIO
```

The Bio.Graphics module depends on the third party Python library ReportLab. Although focused on producing PDF files, ReportLab can also create encapsulated postscript (EPS) and (SVG) files. In addition to these vector-based images, provided certain further dependencies such as the Python Imaging Library (PIL) are installed, ReportLab can also output bitmap images (including JPEG, PNG, GIF, BMP and PICT formats).

3.7.1. GenomeDiagram

The **Bio.Graphics.GenomeDiagram** module was added to Biopython 1.50, having previously been available as a separate Python module dependent on Biopython.

As the name might suggest, GenomeDiagram was designed for drawing whole genomes, in particular prokaryotic genomes, either as linear diagrams (optionally broken up into fragments to fit better) or as circular wheel diagrams. It proved also well suited to drawing quite detailed figures for smaller genomes such as phage, plasmids or mitochrondia.

This module is easiest to use if you have your genome loaded as a SeqRecord object containing lots of SeqFeature objects – for example as loaded from a GenBank file.

3.7.2. Diagrams, Tracks, Feature-Sets, and Features

GenomeDiagram uses a nested set of objects. At the top level, you have a diagram object representing a sequence (or sequence region) along the horizontal axis (or circle). A diagram can contain one or more tracks, shown stacked vertically (or radially on circular diagrams). These will typically all have the same length and represent the same sequence region. You might use one track to show the gene locations, another to show regulatory regions, and a third track to show the GC percentage.

The most commonly used type of track will contain features, bundled together in feature-sets. You might choose to use one feature-set for all your CDS features, and another for tRNA features. This isn't required – they can all go in the same feature-set, but it makes it easier to update the properties of just selected features (e.g., make all the tRNA features red).

There are two main ways to build up a complete diagram. Firstly, the top down approach where you create a diagram object, and then using its methods add track(s), and use the track methods to add feature-set(s), and use their methods to add the features. Secondly, you can create the individual objects separately (in whatever order suits your code), and then combine them.

3.7.3. A Top–Down Example

We're going to draw a whole genome from a SeqRecord object read in from a GenBank file. This example uses the pPCP1 plasmid from Yersinia pestis biovar Microtus (NC_005816.gb)

```
In [3]:
record = SeqIO.read("data/NC_005816.gb", "genbank")
```

We're using a top down approach, so after loading in our sequence we next create an empty diagram, then add an (empty) track, and to that add an (empty) feature set:

```
In [4]:
gd_diagram = GenomeDiagram.Diagram("Yersinia pes-
tis biovar Microtus plasmid pPCP1")
```

```
gd_track_for_features = gd_diagram.new_track(1,
name="Annotated Features")
gd_feature_set = gd_track_for_features.new_set()
```

Now the fun part – we take each gene SeqFeature object in our SeqRecord, and use it to generate a feature on the diagram. We're going to color them blue, alternating between a dark blue and a light blue.

```
In [5]:
for feature in record.features:
     if feature.type != "gene":          #Exclude this
feature
          continue
     if len(gd_feature_set) % 2 == 0:
          color = colors.blue
     else:
          color = colors.lightblue
          gd_feature_set.add_feature(feature,
color=color, label=True)
```

Now we come to actually making the output file. This happens in two steps, first we call the draw method, which creates all the shapes using ReportLab objects. Then we call the write method which renders these to the requested file format. Note you can output in multiple file formats:

```
In [6]:
gd_diagram.draw(format="linear",
orientation="landscape", pagesize='A4',     frag-
ments=4, start=0, end=len(record))
gd_diagram.write("data/plasmid_linear.png", "png")
```

Lets have a look at the previous one:

Notice that the fragments argument which we set to four controls how many pieces the genome gets broken up into.

If you want to do a circular figure, then try this:

```
In [7]:
gd_diagram.draw(format="circular",  circular=True,
pagesize=(20*cm,20*cm),                    start=0,
end=len(record),       circle_core=0.7)   gd_dia-
gram.write("data/plasmid_circular.png",     "PNG")
```

```
Image("data/plasmid_circular.png")
Out[7]:
```

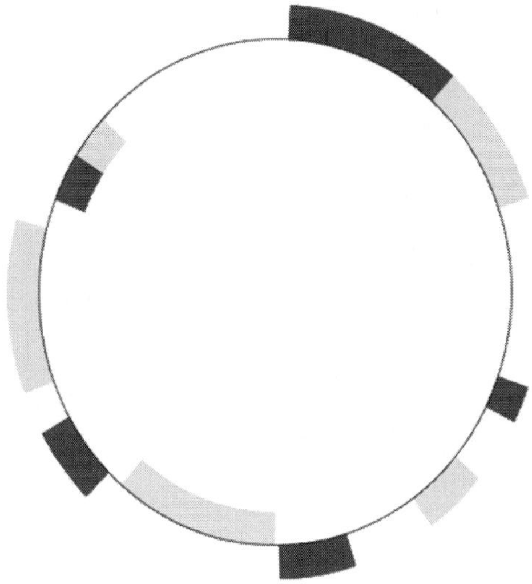

These figures are not very exciting, but we've only just got started.

A bottom up example

Now let's produce exactly the same figures, but using the bottom up approach. This means we create the different objects directly (and this can be done in almost any order) and then combine them.

```
In [8]:
record = SeqIO.read("data/NC_005816.gb", "genbank")
#Create the feature set and its feature objects,
gd_feature_set = GenomeDiagram.FeatureSet()
for feature in record.features:
     if feature.type != "gene":         #Exclude this 
feature
          continue
     if len(gd_feature_set) % 2 == 0:
          color = colors.blue
     else:
          color = colors.lightblue
     gd_feature_set.add_feature(feature, 
color=color, label=True) #(this for loop is the same 
as in the previous example)
#Create a track, and a diagram
```

```
gd_track_for_features = GenomeDiagram.
Track(name="Annotated Features")
gd_diagram = GenomeDiagram.Diagram("Yersinia pes-
tis biovar Microtus plasmid pPCP1")
#Now have to glue the bits together...
gd_track_for_features.add_set(gd_feature_set)
gd_diagram.add_track(gd_track_for_features, 1)
```

You can now call the draw and write methods as before to produce a linear or circular diagram, using the code at the end of the top-down example above. The figures should be identical.

In the above example we used a SeqRecord's SeqFeature objects to build our diagram. Sometimes you won't have SeqFeature objects, but just the coordinates for a feature you want to draw. You have to create minimal SeqFeature object, but this is easy:

```
In [9]:
from Bio.SeqFeature import SeqFeature, FeatureLoca-
tion my_seq_feature = SeqFeature(FeatureLocation(50
,100),strand=+1)
```

For strand, use +1 for the forward strand, -1 for the reverse strand, and None for both. Here is a short self contained example:

```
In [10]:
gdd = GenomeDiagram.Diagram('Test Diagram')
gdt_features = gdd.new_track(1, greytrack=False)
gds_features = gdt_features.new_set() #Add three
features to show the strand options,
feature = SeqFeature(FeatureLocation(25, 125),
strand=+1)
gds_features.add_feature(feature, name="Forward",
label=True)
feature = SeqFeature(FeatureLocation(150, 250),
strand=None)
gds_features.add_feature(feature,
name="Strandless", label=True)
feature = SeqFeature(FeatureLocation(275, 375),
strand=-1)
gds_features.add_feature(feature, name="Reverse",
label=True)
```

```
gdd.draw(format='linear', pagesize=(15*cm,4*cm), fragments=1, start=0, end=400)
gdd.write("data/GD_labels_default.png", "png")
Image("data/GD_labels_default.png")
Out[10]:
```

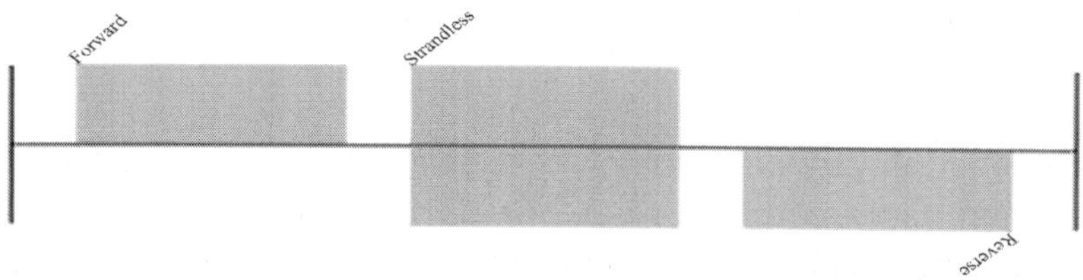

The top part of the image in the next subsection shows the output (in the default feature color, pale green).

Notice that we have used the name argument here to specify the caption text for these features. This is discussed in more detail next.

Recall we used the following (where feature was a SeqFeature object) to add a feature to the diagram:

```
In [11]:
gd_feature_set.add_feature(feature, color=color, label=True)
Out[11]:
<Bio.Graphics.GenomeDiagram._Feature.Feature at 0x7f27bfa2bfd0>
```

In the example above the SeqFeature annotation was used to pick a sensible caption for the features. By default the following possible entries under the SeqFeature object's qualifiers dictionary are used: gene, label, name, locus_tag, and product. More simply, you can specify a name directly:

```
In [12]:
gd_feature_set.add_feature(feature, color=color, label=True, name="My Gene")
Out[12]:
<Bio.Graphics.GenomeDiagram._Feature.Feature at 0x7f27bfa3b048>
```

In addition to the caption text for each feature's label, you can also choose the font, position (this defaults to the start of the sigil, you can also choose the middle or at the end) and orientation (for linear diagrams only, where this defaults to rotated by 45 degrees):

```
In [13]:
#Large font, parallel with the track gd_feature_set.
add_feature(feature, label=True, color="green",
label_size=25, label_angle=0) #Very small font,
perpendicular to the track (towards it) gd_feature_
set.add_feature(feature, label=True, color="purp
le", label_position="e
nd", label_size=4, label_
angle=90) #Small font, perpendicular to the track
(away from it) gd_feature_set.add_feature(feature,
label=True, color="blue", la-
bel_position="middle", la-
bel_size=6, label_angle=-90)
Out[13]:
<Bio.Graphics.GenomeDiagram._Feature.Feature at
0x7f27cc0b2d30>
```

Combining each of these three fragments with the complete example in the previous section should give something like this:

```
In [14]:
gdd.draw(format='linear', pagesize=(15*cm,4*cm),
fragments=1, start=0, end=400) gdd.write("data/
GD_labels.png", "png") Image("data/GD_labels.png")
Out[14]:
```

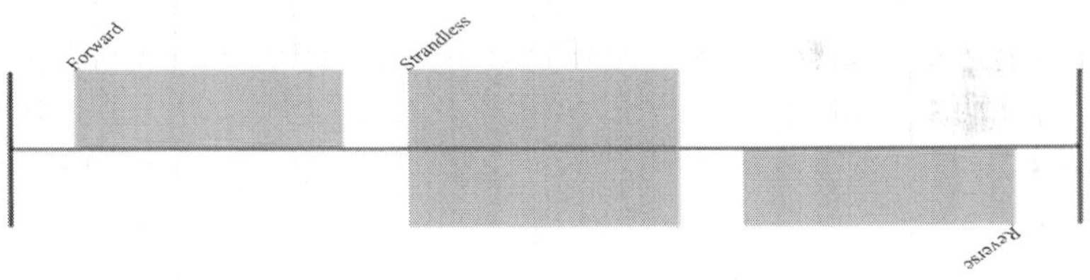

We've not shown it here, but you can also set label_color to control the label's color.

You'll notice the default font is quite small – this makes sense because you will usually be drawing many (small) features on a page, not just a few large ones as shown here.

The examples above have all just used the default sigil for the feature, a plain box, which was all that was available in the last publicly released standalone version of GenomeDiagram. Arrow sigils were included when GenomeDiagram was added to Biopython 1.50:

```
In [15]:
#Default uses a BOX sigil
gd_feature_set.add_feature(feature) #You can make this explicit: gd_feature_set.add_feature(feature, sigil="BOX") #Or opt for an arrow: gd_feature_set.add_feature(feature, sigil="ARROW") #Box with corners cut off (making it an octagon) gd_feature_set.add_feature(feature, sigil="OCTO") #Box with jagged edges (useful for showing breaks in contains) gd_feature_set.add_feature(feature, sigil="JAGGY") #Arrow which spans the axis with strand used only for direction gd_feature_set.add_feature(feature, sigil="BIGARROW")
Out[15]:
<Bio.Graphics.GenomeDiagram._Feature.Feature at 0x7f27bfa3a7f0>
```

These are shown below. Most sigils fit into a bounding box (as given by the default BOX sigil), either above or below the axis for the forward or reverse strand, or straddling it (double the height) for strand-less features. The BIGARROW sigil is different, always straddling the axis with the direction taken from the feature's stand.

We introduced the arrow sigils in the previous section. There are two additional options to adjust the shapes of the arrows, firstly the thickness of the arrow shaft, given as a proportion of the height of the bounding box:

```
In [16]:
#Full height shafts, giving pointed boxes: gd_feature_set.add_feature(feature, sigil="ARROW", color="brown", arrowshaft_height=1.0) #Or, thin shafts: gd_feature_set.add_feature(feature, sigil="ARROW", color="teal", arrowshaft_height=0.2) #Or, very thin shafts: gd_feature_set.add_feature(feature, sigil="ARROW", color="darkgreen", arrowshaft_height=0.1)
Out[16]:
<Bio.Graphics.GenomeDiagram._Feature.Feature at 0x7f27cc0c0400>
```

The results are shown below:

Secondly, the length of the arrow head – given as a proportion of the height of the bounding box (defaulting to 0.5, or 50%):

```
In [17]:
#Short arrow heads: gd_feature_set.add_feature(feature, sigil="ARROW", color="blue", arrowhead_length=0.25) #Or, longer arrow heads: gd_feature_set.add_feature(feature, sigil="ARROW", color="orange", arrow-head_length=1) #Or, very very long arrow heads (i.e. all head, no shaft, so triangles): gd_feature_set.add_feature(feature, sigil="ARROW", color="red", arrowhead_length=10000)
Out[17]:
<Bio.Graphics.GenomeDiagram._Feature.Feature at 0x7f27cc0bc208>
```

The results are shown below:

Biopython 1.61 adds a new BIGARROW sigil which always stradles the axis, pointing left for the reverse strand or right otherwise:

```
In [18]:
#A large arrow straddling the axis: gd_feature_set.add_feature(feature, sigil="BIGARROW")
Out[18]:
<Bio.Graphics.GenomeDiagram._Feature.Feature at 0x7f27bfa2be80>
```

All the shaft and arrow head options shown above for the ARROW sigil can be used for the BIGARROW sigil too.

Now let's return to the pPCP1 plasmid from Yersinia pestis biovar Microtus, and the top down approach used above, but take advantage of the sigil options we've now discussed. This time we'll use arrows for the genes, and overlay them with strand-less features (as plain boxes) showing the position of some restriction digest sites.

```
In [19]:
record = SeqIO.read("data/NC_005816.gb", "gen-bank") gd_diagram = GenomeDiagram.Diagram(record.id) gd_track_for_features = gd_diagram.new_track(1, name="Annotated Features") gd_feature_set = gd_
```

```
track_for_features.new_set() for feature in record.
features:    if feature.type != "gene":         #Ex-
clude this feature            continue     if len(gd_
feature_set) % 2 == 0:                color = colors.
blue       else:               color = colors.light-
blue    gd_feature_set.add_feature(feature, sigil=
"ARROW",                                 color=color,
label=True,                                       la-
bel_size  =  14,  label_angle=0)  #I  want  to  in-
clude  some  strandless  features,  so  for  an  ex-
ample #will  use  EcoRI  recognition  sites  etc. for
site, name, color in [("GAATTC","EcoRI",colors.gre
en),                         ("CCCGGG","SmaI",colors.
orange),                  ("AAGCTT","HindIII",colors.
red),                      ("GGATCC","BamHI",colors.
purple)]:    index = 0    while True:          index
= record.seq.find(site, start=index)        if index
== -1 : break          feature = SeqFeature(Feature
Location(index, index+len(site)))          gd_fea-
ture_set.add_feature(feature, color=color, name=na
me,                                      label=True,
label_size = 10,                                  la-
bel_color=color)          index += len(site) gd_di-
agram.draw(format="linear",  pagesize='A4',  frag-
ments=4,                  start=0, end=len(record))
gd_diagram.write("data/plasmid_linear_nice.png",
"png") Image("data/plasmid_linear_nice.png")
Out[19]:
```

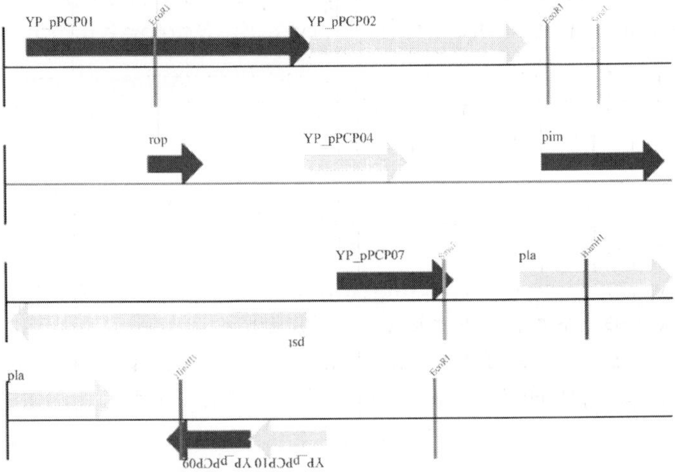


```
In [20]:
gd_diagram.draw(format="circular", circular=True,
pagesize=(20*cm,20*cm), start=0,
end=len(record), circle_core = 0.5) gd_diagram.
write("data/plasmid_circular_nice.png", "png")
Image("data/plasmid_circular_nice.png")
Out[20]:
```

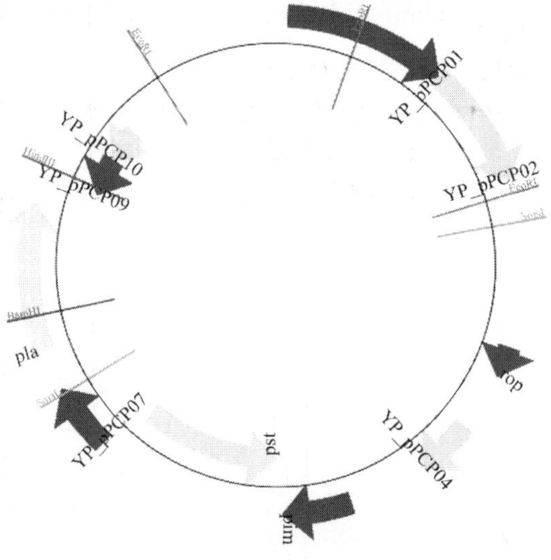

All the examples so far have used a single track, but you can have more than one track – for example show the genes on one, and repeat regions on another. In this example we're going to show three phage genomes side by side to scale. We'll need the GenBank files for the following three phage:

NC_002703 – Lactococcus phage Tuc2009, complete genome (38347 bp)

AF323668 – Bacteriophage bIL285, complete genome (35538 bp)

NC_003212 – Listeria innocua Clip11262, complete genome, of which we are focussing only on integrated prophage 5 (similar length).

You can download these using Entrez if you like. For the third record we've worked out where the phage is integrated into the genome, and slice the record to extract it, and must also reverse complement to match the orientation of the first two phage:

```
In [21]:
A_rec = SeqIO.read("data/NC_002703.gbk", "gb")
B_rec = SeqIO.read("data/AF323668.gbk", "gb")
```

The figure we are imitating used different colors for different gene functions. One way to do this is to edit the GenBank file to record color preferences for each feature – something Sanger's Artemis editor does, and which GenomeDiagram should understand. Here however, we'll just hard code three lists of colors.

Note that the annotation in the GenBank files doesn't exactly match that shown in Proux et al., they have drawn some unannotated genes.

```
In [22]:
from reportlab.lib.colors import red, grey, orange, green, brown, blue, lightblue, purple
A_colors = [red]*5 + [grey]*7 + [orange]*2 + [grey]*2 + [orange] + [grey]*11 + [green]*4 \         + [grey] + [green]*2 + [grey, green] + [brown]*5 + [blue]*4 + [lightblue]*5 \            + [grey, lightblue] + [purple]*2 + [grey] B_colors = [red]*6 + [grey]*8 + [orange]*2 + [grey] + [orange] + [grey]*21 + [green]*5 \          + [grey] + [brown]*4 + [blue]*3 + [lightblue]*3 + [grey]*5 + [purple]*2
```

Now to draw them – this time we add three tracks to the diagram, and also notice they are given different start/end values to reflect their different lengths.

```
In [23]:
name = "data/Proux Fig 6" gd_diagram = GenomeDi-
```

```
agram.Diagram(name) max_len = 0 for record, gene_
colors in zip([A_rec, B_rec], [A_colors, B_col-
ors]): max_len = max(max_len, len(record))
gd_track_for_features = gd_diagram.new_track(1,
name=record.name, gre
ytrack=True, start=0,
end=len(record)) gd_feature_set = gd_track_
for_features.new_set() i = 0 for fea-
ture in record.features: if feature.
type != "gene": #Exclude this feature
continue gd_feature_set.add_feature(feature,
sigil="ARROW", color=gene_
colors[i], label=True, name
= str(i+1), label_po
sition="start", la-
bel_size = 6, label_angle=0) i+=1 gd_di-
agram.draw(format="linear", pagesize='A4', frag-
ments=1, start=0, end=max_len)
gd_diagram.write(name + ."png", "png") Image(name +
."png")
Out[23]:
```

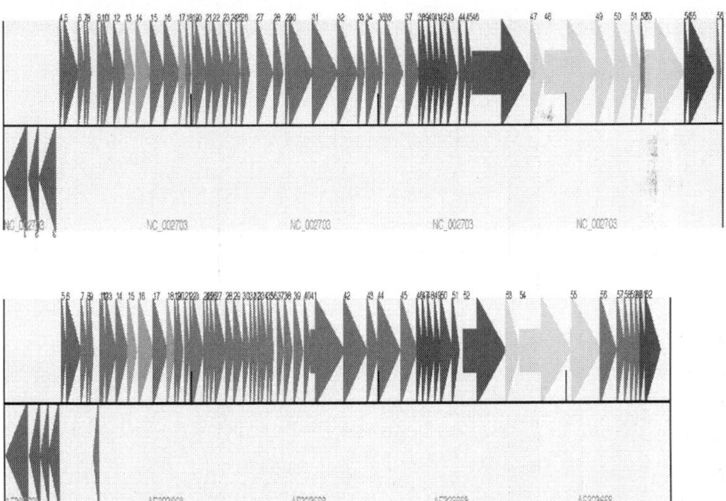

I did wonder why in the original manuscript there were no red or orange genes marked in the bottom phage. Another important point is here the phage are shown with different lengths – this is because they are all drawn to the same scale (they are different lengths).

The key difference from the published figure is they have color-coded links between similar proteins – which is what we will do in the next section.

Biopython 1.59 added the ability to draw cross links between tracks – both simple linear diagrams as we will show here, but also linear diagrams split into fragments and circular diagrams.

Continuing the example from the previous we would need a list of cross links between pairs of genes, along with a score or color to use. Realistically you might extract this from a BLAST file computationally, but here I have manually typed them in.

My naming convention continues to refer to the three phage as A, B and C. Here are the links we want to show between A and B, given as a list of tuples (percentage similarity score, gene in A, gene in B).

```
In [24]:
#Tuc2009 (NC_002703) vs bIL285 (AF323668) A_
vs_B = [ (99, "Tuc2009_01", "int"), (33,
"Tuc2009_03", "orf4"), (94, "Tuc2009_05", "orf6"),
(100,"Tuc2009_06", "orf7"), (97, "Tuc2009_07",
"orf8"), (98, "Tuc2009_08", "orf9"), (98,
"Tuc2009_09", "orf10"), (100,"Tuc2009_10",
"orf12"), (100,"Tuc2009_11", "orf13"), (94,
"Tuc2009_12", "orf14"), (87, "Tuc2009_13",
"orf15"), (94, "Tuc2009_14", "orf16"), (94,
"Tuc2009_15", "orf17"), (88, "Tuc2009_17",
"rusA"), (91, "Tuc2009_18", "orf20"), (93,
"Tuc2009_19", "orf22"), (71, "Tuc2009_20",
"orf23"), (51, "Tuc2009_22", "orf27"), (97,
"Tuc2009_23", "orf28"), (88, "Tuc2009_24",
"orf29"), (26, "Tuc2009_26", "orf38"), (19,
"Tuc2009_46", "orf52"), (77, "Tuc2009_48",
"orf54"), (91, "Tuc2009_49", "orf55"), (95,
"Tuc2009_52", "orf60"), ]
```

For the first and last phage these identifiers are locus tags, for the middle phage there are no locus tags so I've used gene names instead. The following little helper function lets us lookup a feature using either a locus tag or gene name:

```
In [25]:
```

```
def get_feature(features, id, tags=["locus_tag", "gene"]):     """Search list of SeqFeature objects for an identifier under the given tags."""     for f in features:         for key in tags:             #tag may not be present in this feature             for x in f.qualifiers.get(key, []):                 if x == id:                     return f     raise KeyError(id)
```

We can now turn those list of identifier pairs into SeqFeature pairs, and thus find their location co-ordinates. We can now add all that code and the following snippet to the previous example to add cross-links to the figure:

```
In [ ]:
from Bio.Graphics.GenomeDiagram import CrossLink from reportlab.lib import colors #Note it might have been clearer to assign the track numbers explicitly... for rec_X, tn_X, rec_Y, tn_Y, X_vs_Y in [(A_rec, 2, B_rec, 1, A_vs_B)]:     track_X = gd_diagram.tracks[tn_X]     track_Y = gd_diagram.tracks[tn_Y] for score, id_X, id_Y in X_vs_Y:         feature_X = get_feature(rec_X.features, id_X)         feature_Y = get_feature(rec_Y.features, id_Y)         color = colors.linearlyInterpolatedColor(colors.white, colors.firebrick, 0, 100, score)         link_xy = CrossLink((track_X, feature_X.location.start, feature_X.location.end),                 (track_Y, feature_Y.location.start, feature_Y.location.end)                 ,             color, colors.lightgrey) gd_diagram.cross_track_links.append(link_xy)   gd_diagram.draw(format="linear", pagesize='A4', fragments=1,             start=0, end=max_len) gd_diagram.write("data/cross.png", "png") Image("data/cross.png")
Out[ ]:
```

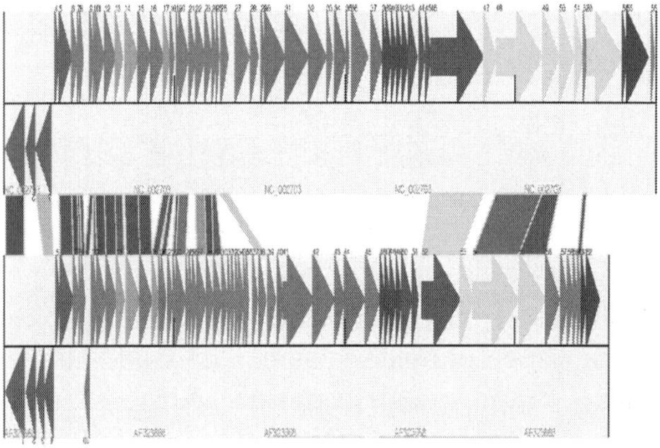

There are several important pieces to this code. First the GenomeDiagram object has a cross_track_links attribute which is just a list of CrossLink objects. Each CrossLink object takes two sets of track-specific co-ordinates (here given as tuples, you can alternatively use a GenomeDiagram.Feature object instead). You can optionally supply a color, border color, and say if this link should be drawn flipped (useful for showing inversions).

You can also see how we turn the BLAST percentage identity score into a color, interpolating between white (0%) and a dark red (100%). In this example we don't have any problems with overlapping cross-links. One way to tackle that is to use transparency in ReportLab, by using colors with their alpha channel set. However, this kind of shaded color scheme combined with overlap transparency would be difficult to interpret. The result:

There is still a lot more that can be done within Biopython to help improve this figure. First of all, the cross links in this case are between proteins which are drawn in a strand specific manor. It can help to add a background region (a feature using the 'BOX' sigil) on the feature track to extend the cross link. Also, we could reduce the vertical height of the feature tracks to allocate more to the links instead – one way to do that is to allocate space for empty tracks. Furthermore, in cases like this where there are no large gene overlaps, we can use the axis-straddling BIGARROW sigil, which allows us to further reduce the vertical space needed for the track.

Beyond that, finishing touches you might want to do manually in a vector image editor include fine-tuning the placement of gene labels, and adding other custom annotation such as highlighting particular regions.

Although not really necessary in this example since none of the cross-links overlap, using a transparent color in ReportLab is a very useful technique for superimposing multiple links. However, in this case a shaded color scheme should be avoided.

3.8. CHROMOSOMES

The Bio.Graphics.BasicChromosome module allows drawing of chromosomes. There is an example in Jupe et al. (2012) using colors to highlight different gene families.

Simple Chromosomes

Important: To continue this example you have first to download a few chromosomes from Arabidopsis thaliana, the code to help you is here:

Very Important: This is slow and clogs the network, you only need to do this once (even if you close the notebook as the download will be persistent)

```
In [ ]:
from ftplib import FTP ftp = FTP('ftp.ncbi.nlm.
nih.gov') print("Logging in") ftp.login() ftp.
cwd('genomes/archive/old_genbank/A_thaliana/OLD/')
print("Starting download - This can be slow!") for
chro, name in [ ("CHR_I", "NC_003070.
fna"), ("CHR_I", "NC_003070.gbk"), ("CHR_
II", "NC_003071.fna"), ("CHR_II", "NC_003071.
gbk"), ("CHR_III", "NC_003072.fna"),
("CHR_III", "NC_003072.gbk"), ("CHR_
IV", "NC_003073.fna"), ("CHR_IV", "NC_003073.
gbk"), ("CHR_V", "NC_003074.fna"), ("CHR_V",
"NC_003074.gbk")]: print("Downloading", chro,
name) ftp.retrbinary('RETR %s/%s' % (chro,
name), open('data/%s' % name, 'wb').write) ftp.
quit() print('Done')
```

Here is a very simple example – for which we'll use Arabidopsis thaliana.

You can skip this bit, but first I downloaded the five sequenced chromosomes from the NCBI's FTP site (per the code above) and then parsed them with Bio.SeqIO to find out their lengths. You could use the GenBank files for this, but it is faster to use the FASTA files for the whole chromosomes:

```
In [ ]:
from Bio import SeqIO entries = [("Chr I", "NC_003070.
```

```
fna"), ("Chr II", "NC_003071.fna"), ("Chr III", "NC_003072.fna"), ("Chr IV", "NC_003073.fna"), ("Chr V", "NC_003074.fna")] for (name, filename) in entries: record = SeqIO.read("data/" + filename, "fasta") print(name, len(record))
```

This gave the lengths of the five chromosomes, which we'll now use in the following short demonstration of the BasicChromosome module:

```
In [ ]:
from reportlab.lib.units import cm from Bio.Graphics import BasicChromosome entries = [("Chr I", 30432563), ("Chr II", 19705359), ("Chr III", 23470805), ("Chr IV", 18585042), ("Chr V", 26992728)] max_len = 30432563 #Could compute this telomere_length = 1000000 #For illustration chr_diagram = BasicChromosome.Organism(output_format="png") chr_diagram.page_size = (29.7*cm, 21*cm) #A4 landscape for name, length in entries: cur_chromosome = BasicChromosome.Chromosome(name) #Set the scale to the MAXIMUM length plus the two telomeres in bp, #want the same scale used on all five chromosomes so they can be #compared to each other cur_chromosome.scale_num = max_len + 2 * telomere_length #Add an opening telomere start = BasicChromosome.TelomereSegment() start.scale = telomere_length cur_chromosome.add(start) #Add a body - using bp as the scale length here. body = BasicChromosome.ChromosomeSegment() body.scale = length cur_chromosome.add(body) #Add a closing telomere end = BasicChromosome.TelomereSegment(inverted=True) end.scale = telomere_length cur_chromosome.add(end) #This chromosome is done chr_diagram.add(cur_chromosome) chr_diagram.draw("data/simple_chrom.png", "Arabidopsis thaliana") Image("data/simple_chrom.png")
```

This example is deliberately short and sweet. The next example shows the location of features of interest.

Continuing from the previous example, let's also show the tRNA genes. We'll get their locations by parsing the GenBank files for the five Arabidopsis thaliana chromosomes. You'll need to download these files from the NCBI FTP site.

```
In [ ]:
entries = [("Chr I", "NC_003070.gbk"), ("Chr II", "NC_003071.gbk"), ("Chr III", "NC_003072.gbk"), ("Chr IV", "NC_003073.gbk"), ("Chr V", "NC_003074.gbk")] max_len = 30432563 #Could compute this telomere_length = 1000000 #For illustration chr_diagram = BasicChromosome.Organism(output_format="png") chr_diagram.page_size = (29.7*cm, 21*cm) #A4 landscape for index, (name, filename) in enumerate(entries): record = SeqIO.read("data/" + filename,"genbank") length = len(record) features = [f for f in record.features if f.type=="tRNA"] #Record an Artemis style integer color in the feature's qualifiers, #1 = Black, 2 = Red, 3 = Green, 4 = blue, 5 =cyan, 6 = purple for f in features: f.qualifiers["color"] = [index+2] cur_chromosome = BasicChromosome.Chromosome(name) #Set the scale to the MAXIMUM length plus the two telomeres in bp, #want the same scale used on all five chromosomes so they can be #compared to each other cur_chromosome.scale_num = max_len + 2 * telomere_length #Add an opening telomere start = BasicChromosome.TelomereSegment() start.scale = telomere_length cur_chromosome.add(start) #Add a body - again using bp as the scale length here. body = BasicChromosome.AnnotatedChromosomeSegment(length, features) body.scale = length cur_chromosome.add(body) #Add a closing telomere end = BasicChromosome.TelomereSegment(inverted=True) end.scale = telomere_length cur_chromosome.add(end) #This chromosome is done chr_diagram.add(cur_chromosome) chr_diagram.draw("data/tRNA_chrom.png", "Arabidopsis thaliana") Image("data/tRNA_chrom.png")
```

3.9. CONCLUSIONS

We have introduced the concept of BLAST in this chapter. The reader should benefit by practicing the examples given in this chapter. With this one can make informed decisions on the sequences which come out of a genomics or a proteomics projects. We have also introduced a way of drawing and understanding the genomic datasets in this chapter. This should help an inquisitive reader to delve deep into genomics.

4 CHAPTER

NATURAL LANGUAGE PROCESSING (NLP)

CONTENTS

Major transformations are taking place in every scientific discipline, and biomedical sciences are no exception. The role of precision medicine has increased dramatically. It now has good potential to make treatments effective. This would be complemented by understanding biological mechanisms and therapeutic effects on patients. It is a matter of concern that all these advancements reach only a small section of patient populations. We illustrate this taking a tumor board: a custom treatment plan is created by a group of highly paid specialists for an individual patient. This involves creating a research report collating information from research literature and different advances relevant to the type of cancer of a patient. If this is really helpful, then probably every cancer patient should benefit from it. Still more than a million people are diagnosed with cancer every year. What this means is that individual and specialist doctors have far larger datasets being generated every second and they have very little time to make sense of it in a comprehensive way.

The obvious question that arises is this:

Can we make use of IT to help patient diagnosis and treatment?

Let use see if this question makes sense in the light of modern IT tools. Certainly with a large collection of intelligent tools today, information retrieval for a given patient profile could be largely automated. This saves considerable time for doctors and helps them to provide potential diagnoses. Treatments could be provided by the experts, and feedback from those suggestions could lead to continual system improvements. This type of setup requires subsystems that draws meaningful information from a broad range of sources from biological and medical phenomena, in addition to diseases and their respective treatments. The biomedical knowledge and literature is stored in textual format which is our natural language (English). This is an amalgamation of scientific literature that explains biological processes and therapeutic mechanisms of action and combines it with the individual patients' journeys our healthcare systems. Hence, there is a need to build systems which can read natural language text and extract biomedical facts. Combining EHRs with drug-protein interactions, can add a new dimension to the diagnosis. The adoption of EHR systems has already helped a large set of patients. There is a consensus that adoption of such systems and subsystems has led to a considerable improvement of healthcare and also the physicians have been benefitted enormously.

Are there any benefits for clinicians?

We need to first understand the limitations of EHR systems. Vendors offering EHR systems comply with the Health Insurance Portability and Accountability Act (HIPAA) of 1996. Let us first understand what HIPAA compliance means in simple terms. HIPAA compliance involves fulfilling the requirements of the Health Insurance Portability and Accountability Act of 1996, its subsequent amendments, and any related legislation such as the Health Information Technology for Economic and Clinical Health (HITECH) Act.

What are the HIPAA compliance requirements?

That question is not so easy to answer as – in places – the requirements of HIPAA are intentionally vague. This is so HIPAA can be applied equally to every different type of Covered Entity or Business Associate that comes into contact with Protected Health Information (PHI).

EHR systems are designed to comply with HIPAA compliance list. This list is ever-increasing and vendors strive to comply. The list has special focus, covering areas like

1. patient data safety features,
2. reporting requirements, and
3. quality measures.

All these are geared towards innovative functionality that may significantly improve the clinician outcomes and staff efficiency. A major roadblock in this process is access to external data in a coherent fashion. This prevents generating insights, actionable, from all the data (unstructured and/or structured) within the EHR/EMR data.

What are the problems, which one (clinicians) would face?

We list a few of the previously reported common problems with data extraction below:

1. The main reason for coding the data is for billing purposes.
2. Any data available in a free text format, is not available in usable format.
3. As the data in EHRs are not normalized, data from multiple sources against a particular indicator, may have a lot of inconsistencies.

4. Depending on institutional regulations, data may have a problem of incompleteness due to low filling rates.
5. And the most common problem is falsified data in EHR.

With these issues introduced, natural language processing can serve as a layering stone to solving the issues. This can help in transitioning to more intelligent, intuitive, and streamlined reports which medical professionals can use on a regular basis.

4.1. WHAT IS NLP?

Natural language processing (NLP) can be thought of utilizing computer algorithms to process and extract meaning from the natural language like speech or written text. Medical NLP systems specifically can help coping with the next set of common tasks:

Locating, extracting, and summarizing key concepts or phrases from blocks of narrative texts (e.g., clinical notes or a patient's account) for further analysis. Many reports of NLP programs helping in improving medical diagnosis are common. Cases abound where identifying at-risk patients with heart failure with higher accuracy are reported based on a NLP based program. Basically, the program analyzed the data within EMR and notifies medical professionals that these types of patients are at the hospital and need additional attention.

1. It assists in converting important medical data from computer based formats (machine-readable) into natural language (speech or written text).
2. One of the important scope of medical NLP is to optimize unstructured text data. Mapping it to the respective fields greatly improves an EHR.
3. Speech recognition and instant conversion to text is another area which fall within the scope of NLP. Doing speech recognition tasks will allow users to dictate notes. These can then be instantly turned into text. The accuracy of such tasks has increased considerably.
4. NLP plays an important role to assist physicians who spend a great deal of time on paperwork with inputting/organizing data within the EHR system. Automating such routines can significantly improve the staff's productivity, efficiency, and morale.

Careful design and implementation of NLP apps can drastically reduce the time needed to uncover valuable insights from all the data at hand. This would also make physicians' work smart and intelligent. The broad goal of such NLP-powered context engines is to mine unstructured documents at high speed and deliver actionable suggestions which can enhance clinical decision support, sepsis management, stroke management and so on.

The particular allure of NLP for healthcare is that these systems can "learn" over time and deliver better and better insights as more interactions occur. Data processing can be also improved over time by adjusting the statistical probabilities, which account for the likelihood that the delivered bit of data matches the user request. Based on the feedback, these probabilities can be tweaked further to improve the results even further.

4.1.1. Key NLP Use Cases in Healthcare

Let use see how NLP can play a role in healthcare in this section. NLP roles in healthcare is projected to increase and by 2024 it would be a billion-dollar technology enterprise in itself. It is clear as to why there is a rise in demand in using NLP in healthcare, mainly due to the increased data flow. Undoubtedly the available information to the clinicians is beneficial for the industry, but extracting relevant bits of information is still a challenge.

A few innovative NLP based companies have emerged during the past few years. These are pushing towards addressing the common problems and emerging challenges in healthcare data management which also involves providing clues to improved healthcare. This stage passes through processing of unstructured data, computational terminology, which is itself a very challenging task. One can think in terms of PDF lab reports, EHR inputs, and soon.

Majority of the instances involve doctor's handwritten notes, which could be hard to use repeatedly. This just results in a pile-up of such notes. An NLP based system, in this instance, can reduce the time required to extract required information and comprehend valuable insights. We highlight below a few benefits of implementing such a system:

1. Provide a complete, easy-to-review, summary of all the information available to the patient.
2. Alert about the possible risks and chronic conditions the person has.

3. Summarize the key phrases and cluster them into groups such as "heart failure risk" to support predictive modeling.
4. Merge structured and unstructured data to present a detailed overview of the patient's health over time and all previous medical interventions.
5. Apply clinical best practice models to assess compliance with good practice.
6. Expedite Patient Identification for Clinical Trial Enrollment (CTE).
7. Healthcare NLP can also enable finding of the best candidates for clinical trials and medical research. After collecting a large set of potential candidates, it evaluates the information suggesting a possible hit case of a suitable candidate for enrollment.

This can also provide additional information if a certain patient can be asked to participate in a certain clinical trial, while still not discharged. Although, this involves taking consent of the patient for participation still it saves the loss of a suitable candidates. In this way, the patient is also benefitted as he is informed of all the possible treatment options. These are crucial during the initial phases of diseases. An exercise of above steps has been seen to be promising. This drastically improves the trial screening efficiency.

4.1.2. Current Trends in Biological NLP

The core tasks of Biological NLP (BioNLP) are geared towards fine-grained information extraction (IE). There is a forum which specifically addresses the rising and upcoming trends and themes in BioNLP called BioNLP Shared Task (BioNLP-ST). Basically, it addresses the following:

Identification of biologically relevant extraction targets and propose a linguistically motivated approach to event representation

The task starts with manually annotated data being provided for training, development and evaluation of information extraction methods. According to their relevance for biological studies, the annotations are either bound to specific expressions in the text or represented as structured knowledge. Many tools for the detailed evaluation and graphical visualization of annotations and system outputs will be available for participants. Support in performing linguistic processing will be provided to the participants in the form of analyzes created by various state-of-the art tools on the dataset texts. It is advised that a reader interested in employing NLP in biology, not necessarily healthcare or biomedical sciences, participate in such a forum and benefit from the proceedings.

4.2. SIMPLE CLINICAL NATURAL LANGUAGE PROCESSING WITH PYCONTEXTNLP

In this section we introduce the basics of pyConTextNLP, a simple Python tool that we have used extensively for processing clinical text, including radiology, psychiatry, etc.

pyConTextNLP is built around the concept of **targets** and **modifiers**:

- the *target* is the concept we are interested in identifying (like a cough or a pulmonary embolism);
- a *modifier* is a concept that changes the target in some sense (e.g., historical, severity, certainty, negation).

pyConTextNLP relies on regular expressions to identify concepts (both targets and modifiers) within a sentence and then uses simple lexical rules to assign relationships between the identified targets and modifiers. Internally, pyConTextNLP uses graphs. Targets and modifiers are nodes in the graph and relationships between modifiers and targets are edges in the graph.

4.3. SPECIFYING TARGETS, MODIFIERS, AND RULES

pyConTextNLP uses a four-tuple to represent concepts. Within the program we create an instance of an itemData class. Each itemData consists of the following four attributes:

- A **literal** (e.g., "pulmonary embolism", "no definite evidence of"): This is a lingustic representation of the target or modifier we want to identify
- A **category** (e.g., "CRITICAL_FINDING", "PROBABLE_EXISTENCE"): This is the label we want applied to the literal when we see it in text
- A **regular expression** that defines how to identify the literal concept. If no regular expression is specified, a regular expression will be built directly from the literal by wrapping it with word boundaries (e.g., r"""\bpulmonary embolism\b""")

- A **rule** that defines how the concept works in the sentence (e.g., a negation term that looks **forward** in the sentence). This only applies to modifiers.

```
In [ ]:
!pip install -U pycontextnlp==0.6.1.1
In [ ]:
import pyConTextNLP.pyConTextGraph as pyConText
import pyConTextNLP.itemData as itemData
```

We would like to identify patients with pulmonary embolism from radiology reports.

How is the concept of pulmonary embolism represented in the reports – fill in the list below with literals you want to use.

```
In [ ]:
mytargets = itemData.itemData() mytargets.
extend([["pulmonary embolism", "CRITICAL_FINDING",
"", ""], ["pneumonia", "CRITI-
CAL_FINDING", "", ""]])
In [ ]:
print(mytargets)
In [ ]:
!pip install -U radnlp==0.2.0.8
```

4.3.1. Sentence Splitting

pyConTextNLP operates on a *sentence* level and so the first step we need to take is to split our document into individual sentences. pyConTextNLP comes with a simple sentence splitter class.

```
In [ ]:
import pyConTextNLP.helpers as helpers
spliter = helpers.sentenceSplitter()
spliter.splitSentences("This is Dr. Chapman's first
sentence. This is the 2.0 sentence.")
```

However, sentence splitting is a common NLP task and so most full-fledged NLP applications provide sentence splitters. We usually rely on the sentence splitter that is part of the TextBlob package, which in turn relies on the Natural Language Toolkit (NLTK). So before proceeding we need to download some NLTK resources with the following command.

```
In [ ]:
!python -m textblob.download_corpora
```

4.3.2. Demonstration of Basic Sentence Markup With pyConTextNLP

pyConTextNLP uses NetworkX directional graphs to represent the markup: nodes in the graph will be the concepts that are identified in the sentence and edges in the graph will be the relationships between those concepts.

```
In [1]:
import pyConTextNLP.pyConTextGraph as pyConText
import pyConTextNLP.itemData as itemData
import networkx as nx
```

- pyConTextGraph contains the bulk of the pyConTextNLP functionality, including basic class definitions such as the ConTextMarkup class that represents the markup of a sentence.
- itemData contains a class definition for an itemData and functions for reading itemData definitions which are assumed to be in a tab separated file that is specified as either a local file or a remote resource. In this example we will read definitions straight from the GitHub repository.
- An itemData in its most basic form is a four-tuple consisting of
- A **literal** (e.g., "pulmonary embolism", "no definite evidence of")
- A **category** (e.g., "CRITICAL_FINDING", "PROBABLE_EXISTENCE")
- A **regular expression** that defines how to identify the literal concept. If no regular expression is specified, a regular expression will be built directly from the literal by wrapping it with word boundaries (e.g., r"""""\bpulmonary embolism\b""""")
- A **rule** that defines how the concept works in the sentence (e.g., a negation term that looks **forward** in the sentence).

4.3.3. Sentences

These example reports are taken from (with modification) the MIMIC2 demo data set that is a publically available database of de-identified medical records for deceased individuals.

```
In [2]:
reports = """IMPRESSION: Evaluation limited by lack of IV contrast; however, no evidence of bowel obstruction or mass identified within the abdomen or pelvis. Non-specific interstitial opacities and bronchiectasis seen at the right base, suggestive of post-inflammatory changes.""", """IMPRESSION: Evidence of early pulmonary vascular congestion and interstitial edema. Probable scarring at the medial aspect of the right lung base, with no definite consolidation.""" , """IMPRESSION: 1. 2.0 cm cyst of the right renal lower pole. Otherwise, normal appearance of the right kidney with patent vasculature and no sonographic evidence of renal artery stenosis. 2. Surgically absent left kidney.""", """IMPRESSION: No pneumothorax.""", """IMPRESSION: No definite pneumothorax""" """IMPRESSION: New opacity at the left lower lobe consistent with pneumonia.""" ].
```

4.3.4. Read the itemData Definitions

We're reading directly from GitHub. You could read from a local file using a file:// URL.

```
In [3]:
modifiers = itemData.instantiateFromCSVtoitemData( "https://raw.githubusercontent.com/chapmanbe/pyConTextNLP/master/KB/lexical_kb_05042016.tsv")
targets = itemData.instantiateFromCSVtoitemData( "https://raw.githubusercontent.com/chapmanbe/pyConTextNLP/master/KB/utah_crit.tsv")
```

4.3.5. Example Function to Analyze Each Sentence

This the function we'll use for each report. The following section of this document steps through each line.

```
In [4]:
def markup_sentence(s, modifiers, targets, prune_inactive=True):
```

```
"""
"""
markup = pyConText.ConTextMarkup()
markup.setRawText(s)
markup.cleanText()
markup.markItems(modifiers, mode="modifier")
markup.markItems(targets, mode="target")
markup.pruneMarks()
markup.dropMarks('Exclusion')      # apply modifiers
to any targets within the modifiers scope
markup.applyModifiers()
markup.pruneSelfModifyingRelationships()
if prune_inactive:
     markup.dropInactiveModifiers()
return markup
```

We're going to start with our simplest of sentences

```
In [5]:
reports[3]
Out[5]:
'IMPRESSION:  No pneumothorax.'
marking up a sentence
```

We start by creating an instance of the ConTextMarkup class. This is a subclass of a NetworkX DiGraph. Information will be stored in the nodes and edges.

```
In [6]:
markup = pyConText.ConTextMarkup()
In [7]:
isinstance(markup,nx.DiGraph)
Out[7]:
True
In [8]:
#### Set the text to be processed
In [9]:
markup.setRawText(reports[3].lower()) print(markup)
print(len(markup.getRawText()))
__________________________________________________________ raw-
Text:  impression:   no  pneumothorax. cleanedText:
```

```
None ____________________________________________ 29
```

4.3.6. Clean the Text

Prior to processing we do some basic cleaning of the text, such as replacing multiple white spaces with a single space. You'll notice this in the spacing between the colon and "no" in the raw and clean versions of the text.

```
In [10]:
markup.cleanText() print(markup) print(len(markup.
getText()))
____________________________________________ raw-
Text: impression:  no pneumothorax. cleanedText:
impression: no pneumothorax. ____________________
____________________ 28
```

4.3.7. Identify Concepts in the Sentence

The markItems method takes a list of itemData and uses the regular expressions to identify any instances of the itemData in the sentence. With the mode keyword we specify whether these itemData are targets or modifiers. This value will be stored as a data attribute of the node that is created in the graph for any identified concepts.

```
In [11]:
markup.markItems(modifiers, mode="modifier")
print(markup.nodes(data=True))
print(type(markup.nodes()[0]))
[(<id> 215455046329351668380259283016206441424 </
id> <phrase> no </phrase> <category> ['definite_ne-
gated_existence'] </category> , {'category': 'modi-
fier'})]
<class 'pyConTextNLP.pyConTextGraph.tagObject'>
```

What does our initial markup look like?

- We've identified one concept in the sentence: no
- We've created a tagObject for this concept which keeps track of the actual phrase identified by the regular expression, what the category of the itemData was (definite_negated_existence), this is a list because there can be multiple categories. There is also an absurdly long identifier for the node. Note that our mode modifier has been stored as a data element of the node. In NetworkX each node (or edge) has a dictionary for data.

Now let's markup the targets

```
In [12]:
markup.markItems(targets, mode="target")
In [13]:
print(markup.nodes(data=True))
[(<id> 215496472358685501776852452868303018960 </
id> <phrase> pneumothorax </phrase> <category>
['pneumothorax'] </category> , {'category': 'tar-
get'}), (<id> 215455046329351668380259283016206441
424 </id> <phrase> no </phrase> <category> ['defi-
nite_negated_existence'] </category> , {'category':
'modifier'})]
```

What does our markup look like now?

We've added another node to the graph. This time the target pneumothorax.

4.3.8. Prune Marks

After identifying concepts, we prune concepts that are a subset of another identified concept. This results in no changes here, but the importance will be shown later with a different sentence.

```
In [14]:
markup.pruneMarks() print(markup.nodes())
[<id> 215496472358685501776852452868303018960 </
id> <phrase> pneumothorax </phrase> <category>
['pneumothorax'] </category> , <id> 21545504632
9351668380259283016206441424 </id> <phrase> no </
phrase> <category> ['definite_negated_existence']
</category> ]
```

Are there any relationships in our markup?

We do not yet have any relationships (edges) between our concepts (target and modifier edges)

```
In [15]:
print(markup.edges())
[]
```

4.3.9. Apply Modifiers

We now call the applyModifiers method of the ConTextMarkup object to identify any relationships between the nodes.

```
In [16]:
markup.applyModifiers()
print(markup.edges())
[(<id> 215455046329351668380259283016206441424 </id> <phrase> no </phrase> <category> ['definite_negated_existence'] </category> , <id> 215496472358685501776852452868303018960 </id> <phrase> pneumothorax </phrase> <category> ['pneumothorax'] </category> )]
```

We now have a relationship!

We now have a directed edge between our no node and our pneumothorax node. This will be interpreted as pneumothorax being a definitely negated concept in the sentence.

An ever-so-slightly more complex sentence

Let's use a slightly more complex sentence that will illustrate pruning.

```
In [1]:
import pyConTextNLP.pyConTextGraph as pyConText
import pyConTextNLP.itemData as itemData
import networkx as nx
```

Sentences

These example reports are taken from (with modification) the MIMIC2 demo data set that is a publically available database of de-identified medical records for deceased individuals.

```
In [2]:
reports = [ """IMPRESSION: Evaluation limited by lack of IV contrast; however, no evidence of bowel obstruction or mass identified within the abdomen or pelvis. Non-specific interstitial opacities and bronchiectasis seen at the right base, suggestive of post-inflammatory changes.""", """IMPRESSION: Evidence of early pulmonary vascular congestion and interstitial edema. Probable scarring at the medial aspect of the right lung base, with no definite consolidation.""" , """IMPRESSION:
```

```
1. 2.0 cm cyst of the right renal lower pole. Oth-
erwise, normal appearance    of the right kidney
with patent vasculature and no sonographic evidence
of     renal artery stenosis.     2. Surgically
absent left kidney.""",    """IMPRESSION: No pneu-
mothorax.""",    """IMPRESSION: No definite pneumo-
thorax""",    """IMPRESSION:  New opacity at the
left lower lobe consistent with pneumonia.""" ]
```

Read the itemData definitions

```
In [3]:
modifiers  =  itemData.instantiateFromCSVtoitem-
Data(     "https://raw.githubusercontent.com/
chapmanbe/pyConTextNLP/master/KB/lexical_
kb_05042016.tsv") targets = itemData.instantiate-
FromCSVtoitemData(     "https://raw.githubuser-
content.com/chapmanbe/pyConTextNLP/master/KB/
utah_crit.tsv")
```

We're going to start with our simplest of sentences

```
In [4]:
reports[4]
Out[4]:
'IMPRESSION: No definite pneumothorax'.
```

4.3.10. Marking Up a Sentence

We start by creating an instance of the ConTextMarkup class. This is a subclass of a NetworkX DiGraph. Information will be stored in the nodes and edges.

```
In [5]:
markup = pyConText.ConTextMarkup()
In [6]:
markup.setRawText(reports[4].lower())
print(markup)
print(len(markup.getRawText()))
markup.cleanText()
print(markup)
print(len(markup.getText()))
____________________________________________ raw-
```

```
Text: impression: no definite pneumothorax cleaned-
Text:   None ______________________________________
______   36 _______________________________________
___ rawText: impression: no definite pneumothorax
cleanedText: impression: no definite pneumothorax __
_________________________________________ 36
```

Identify concepts in the sentence

```
In [7]:
markup.markItems(modifiers, mode="modifier")
markup.markItems(targets, mode="target")
print(markup.nodes(data=True))
[(<id> 114731505145955200893208409019887573968 </
id> <phrase> pneumothorax </phrase> <category>
['pneumothorax'] </category> , {'category': 'tar-
get'}), (<id> 1147078358908603894536448157481345259
904 </id> <phrase> no </phrase> <category> ['defi-
nite_negated_existence'] </category> , {'category':
'modifier'}), (<id> 114696347015014195979537648822800276432 </id> <phrase> definite </phrase> <catego-
ry> ['definite_existence'] </category> , {'catego-
ry': 'modifier'}), (<id> 1147081860793387024994806
08197255289808 </id> <phrase> no definite </phrase>
<category> ['definite_negated_existence'] </catego-
ry> , {'category': 'modifier'})]
```

What does our initial markup look like?

We've identified three concepts in the sentence:

- "no"
- "no definite"
- "pneumothorax"

Here "no" is not a true concept in the sentence; it is a subset of the concept "no definite"

Prune Marks

After identifying concepts, we prune concepts that are a subset of another identified concept.

```
In [8]:
markup.pruneMarks() print(markup.nodes())
[<id> 114731505145955200893208409019887573968 </
```

```
id> <phrase> pneumothorax </phrase> <category>
['pneumothorax'] </category> , <id> 11470818607933
870249948060819725528980 8 </id> <phrase> no definite
</phrase> <category> ['definite_negated_existence']
</category> ]
```

What is the effect of pruneMarks

We've correctly dropped no as an identified concept.

Apply modifiers

We now call the applyModifiers method of the ConTextMarkup object to identify any relationships between the nodes.

```
In [9]:
markup.applyModifiers()
print(markup.edges())
[(<id> 114708186079338702499480608197255289808 </
id> <phrase> no definite </phrase> <category> ['defi-
nite_negated_existence'] </category> , <id> 1147
31505145955200893208409019887573968 </id> <phrase>
pneumothorax </phrase> <category> ['pneumothorax']
</category> )]
```

4.4. PROCESSING MULTISENTENCE DOCUMENTS

```
In [1]:
import pyConTextNLP.pyConTextGraph as pyConText
import pyConTextNLP.itemData as itemData
from textblob import TextBlob
import networkx as nx
import pyConTextNLP.display.html as html
from IPython.display import display, HTML
In [2]:
reports = [     """IMPRESSION: Evaluation limited
by lack of IV contrast; however, no evidence of
bowel obstruction or mass identified within the ab-
domen or pelvis. Non-specific interstitial opacities
and bronchiectasis seen at the right    base, sug-
gestive of post-inflammatory changes.""",    """IM-
PRESSION: Evidence of early pulmonary vascular con-
gestion and interstitial edema. Probable scarring
```

```
at the medial aspect of the right lung base, with no definite consolidation.""" , """IMPRESSION: 1. 2.0 cm cyst of the right renal lower pole. Otherwise, normal appearance of the right kidney with patent vasculature and no sonographic evidence of renal artery stenosis. 2. Surgically absent left kidney.""", """IMPRESSION: No pneumothorax.""", """IMPRESSION: No definite pneumothorax""" """IMPRESSION: New opacity at the left lower lobe consistent with pneumonia.""" ]
In [3]:
modifiers = itemData.instantiateFromCSVtoitemData( "https://raw.githubusercontent.com/chapmanbe/pyConTextNLP/master/KB/lexical_kb_05042016.tsv")
targets = itemData.instantiateFromCSVtoitemData( "https://raw.githubusercontent.com/chapmanbe/pyConTextNLP/master/KB/utah_crit.tsv")
```

4.4.1. Define Markup_Sentence

We are putting the functionality we went through in the previous two notebooks (BasicSentenceMarkup and BasicSentenceMarkupPart2) into a function markup_sentence. We add one step to the function: dropInactiveModifiers will delete any modifier node that does not get attached to a target node.

```
In [4]:
def markup_sentence(s, modifiers, targets, prune_inactive=True):
"""
"""
markup = pyConText.ConTextMarkup()
markup.setRawText(s)
markup.cleanText()
markup.markItems(modifiers, mode="modifier")
markup.markItems(targets, mode="target")
markup.pruneMarks()
markup.dropMarks('Exclusion')     # apply modifiers to any targets within the modifiers scope
markup.applyModifiers()
markup.pruneSelfModifyingRelationships()
```

```
if prune_inactive:
     markup.dropInactiveModifiers()
return markup
In [5]:
report = reports[0]
print(report)
IMPRESSION: Evaluation limited by lack of IV con-
trast; however, no evidence of      bowel obstruc-
tion or mass identified within the abdomen or pelvis.
Non-specific interstitial opacities and bronchiec-
tasis seen at the right base, suggestive of post-
inflammatory changes.
```

4.4.2. Create a ConTextDocument

ConTextDocument is a class for organizing the markup of multiple sentences. It has a private attribute that is NetworkX DiGraph that represents the document structure. In this example, we only use the ConTextDocument class to collect multiple sentence markups.

```
In [6]:
context = pyConText.ConTextDocument()
```

Split the document into sentences and process each sentence

pyConTextNLP comes with a simple sentence splitter in helper.py. I have not been maintaining this and have recently been using TextBlob to split sentences. A known problem with either sentence splitting solution is enumerated lists that don't use periods.

```
In [7]:
blob = TextBlob(report.lower())
count = 0
rslts = []
for s in blob.sentences:
     m = markup_sentence(s.raw, modifiers=modifiers,
targets=targets)     rslts.append(m)
for r in rslts:
     context.addMarkup(r)
```

4.4.3. Displaying pyConTextNLP Markups

The display subpackage contains some functionality for visualizing the markups. Here I use HTML to color-code identified concepts.

```
In [8]:
clrs = {\    "bowel_obstruction": "blue",    "in-
flammation":  "blue",        "definite_negated_exis-
tence": "red",      "probable_negated_existence":
"indianred",     "ambivalent_existence": "orange",
"probable_existence": "forestgreen",     "definite_
existence": "green",    "historical": "goldenrod",
"indication": "pink",     "acute": "golden" }
In [9]:
display(HTML(html.mark_document_with_
html(context,colors=clrs, default_color="black")))
impression: evaluation limited by lack of iv con-
trast; however, no evidence of bowel obstruction or
mass identified within the abdomen or pelvis. non-
specific interstitial opacities and bronchiectasis
seen at the right base, suggestive of post-inflamma-
tory changes.
There is also a rich XML description of the Con-
TextDocument
In [10]:
print(context.getXML())
```

4.5. CLASSIFYING DOCUMENTS

In this section we demonstrate a basic document level classification of reports with respect to a single finding (fever). We leverage the convenience of Pandas to read our data from a SQLite database and then use Pandas to add our classification as a new column in the dataframe.

Many of the common pyConTextNLP tasks have been wrapped into functions contained in the radnlp pacakge. We important multiple modules that will allow us to write concise code.

```
In [1]:
import pyConTextNLP.pyConTextGraph as pyConText
import pyConTextNLP.itemData as itemData
import pymysql
import numpy as np
import os
import radnlp.io  as rio
```

```
import radnlp.view as rview
import radnlp.rules as rules
import radnlp.schema as schema
import radnlp.utils as utils
import radnlp.split as split
import radnlp.classifier as classifier
import pandas as pd from IPython.display
import clear_output, display, HTML, Image
from IPython.html.widgets import interact, interac-
tive, fixed
from IPython.display import clear_output
import ipywidgets as widgets from radnlp.data
import classrslts
import networkx as nx
conn = pymysql.connect(host="mys
ql", port=3306,user="jovyan",
passwd='jovyan',db='mimic2')
/opt/conda/lib/python3.5/site-packages/IPython/
html.py:14: ShimWarning: The `IPython.html` package
has been deprecated. You should import from `note-
book` instead. `IPython.html.widgets` has moved to
`ipywidgets`. "`IPython.html.widgets` has moved
to `ipywidgets`.", ShimWarning)
In [2]:
conn = pymysql.connect(host="mys
ql", port=3306,user="jovyan",
passwd='jovyan',db='mimic2')
In [63]:
colors={"pulmonary_embolism":"blue",
"pneumonia":"blue", "pneumothorax":"blue",
"diverticulitis":"blue", "definite_ne-
gated_existence":"red", "probable_negat-
ed_existence":"indianred", "ambivalent_
existence":"orange", "probable_existence":"
forestgreen", "definite_existence":"green",
"historical":"goldenrod", "indication":"Pink",
"acute":"golden"}
In [4]:
import radnlp radnlp.__version__
```

```
Out[4]:
'0.2.0.8'
```

4.5.1. Explanation of getOptions

This is just kind of a port of a command line application where I'd use argparse to get the options.

```
In [51]:
def getOptions():    """Generates arguments for specifying database and other parameters"""    options = {}    options['lexical_kb'] = ["https://raw.githubusercontent.com/chapmanbe/pyConTextNLP/master/KB/lexical_kb_05042016.tsv"]    options['domain_kb'] = ["https://raw.githubusercontent.com/chapmanbe/pyConTextNLP/master/KB/critical_findings.tsv"]    options["schema"] = "https://raw.githubusercontent.com/chapmanbe/pyConTextNLP/master/KB/schema2.csv"    options["rules"] = "https://raw.githubusercontent.com/chapmanbe/pyConTextNLP/master/KB/classificationRules3.csv"    return options
```

4.5.2. Explanation of get_kb_rules_schema

itemData.instantiateFromCSVtoitemData: This function is a somewhat unfortunate name as it implies reading CSV files but we've moved to tab delimited files, since we sometimes have commas in the regular expressions.

```
In [52]:
def get_kb_rules_schema(options):
"""    Get the relevant kb, rules, and schema.
"""

_radnlp_rules = rules.read_rules(options["rules"])
_schema = schema.read_schema(options["schema"])
modifiers = itemData.itemData()
targets = itemData.itemData()
for kb in options['lexical_kb']:
     modifiers.extend( itemData.instantiateFromCSVtoitemData(kb) )
     for kb in options['domain_kb']:
          targets.extend( itemData.instantiateFromCSVtoitemData(kb) )     return {"rules":_radnlp_
```

```
rules,          "schema":_schema,               "mo
difiers":modifiers,            "targets":targets}
In [53]:
def  analyze_report(report,  modifiers,  targets,
rules, schema):
"""     given an individual radiology report, cre-
ates a pyConTextGraph    object that contains the
context markup    report: a text string containing
the radiology reports     """
markup   =  utils.mark_report(split.get_sentences(
report),                                modifiers,
targets)
clssfy =   classifier.classify_document_targets(mar
kup,                                          rul
es[0],                                        rul
es[1],                                        rul
es[2],                                     schema)
     return    classrslts(context_document=markup,
exam_type="ctpa",  report_text=report,  classifica-
tion_result=clssfy)
In [55]:
options = getOptions()
kb = get_kb_rules_schema(options)
#data = data.dropna()
In [56]:
data = pd.read_sql("""
SELECT noteevents.subject_id,                  no-
teevents.text,                        icd9.code
FROM noteevents
INNER JOIN icd9
ON                         noteevents.subject_id =
icd9.subject_id                     WHERE (   icd9.
code LIKE '415.1%'                                )
AND  noteevents.category  =  'RADIOLOGY_REPORT'""",
conn).drop_duplicates() data.head(10)
Out[56]:

In [57]:
doc_split=["IMPRESSION:", "INTERPRETATION:", "CON-
```

```
CLUSION:"]
def find_impression(text, split):
     for term in split:
          if term in text:
          return text.split(term)[1]
     return np.NaN
data["impression"] = data.apply(lambda row: find_
impression(row["text"], doc_split), axis=1)
data = data.dropna(axis=0, inplace=False)
data["chest"] = data.apply(lambda x: 'chest' in
x["text"].lower() and 'ct' in x["text"].lower(),
axis=1)
data = data[data["chest"] == True]
data = data.reset_index()
print(data.shape)
data.head()
Out[57]:
```

4.6. DOCUMENT CLASSIFICATION

We now need to apply our schema to the reports. Since our data is in a Pandas data frame, the easiest way to process our reports is with the DataFrame apply method.

- We use lambda to create an anonymous function which basically just applies analyze_report to the "impression" column with the modifiers, targets, etc. that we have read in separately.
- analyze_report returns a dictionary with keys as any identified targets defined in the "targets" file and values as a tuple with values:

The schema value that was selected for the document

The node (evidence) that was used for selecting that schema value

```
In [58]:
data["pe rslt"] = data.apply(lambda x: analyze_repo
rt(x["impression"],                              kb
["modifiers"],                                  kb[
"targets"],                                  kb["ru
les"],                              kb["schema"]),
axis=1)
data.head()
```

```
Out[58]:

In [59]:
def view_markup(reports, colors):
     @interact(i=widgets.IntSlider(min=0,
max=len(reports)-1))
def _view_markup(i):
     markup = reports[“pe rslt”][i]
     rview.markup_to_pydot(markup)
     display(Image(“tmp.png”))
mt = rview.markup_to_html(markup, color_map=colors)
display(HTML(mt))
In [64]:
view_markup(data, colors)
```

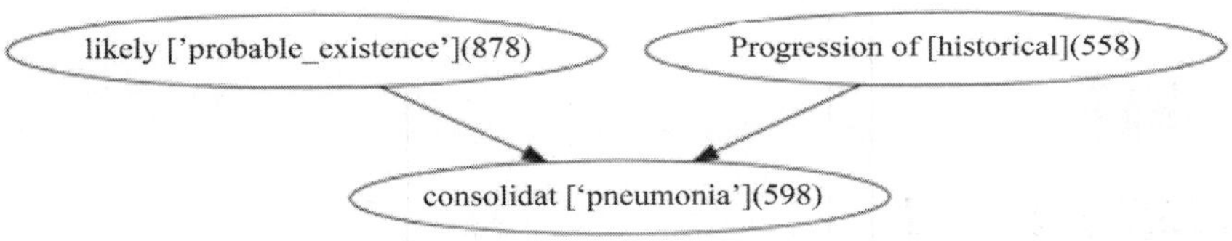

PE Finder Case Review

Report	Classification

Progression of ground-glass opacities to consolidation predominantly in the upper lobes, likely toxicity or hypersensitivity to recently instituted medication. drug-related toxicity. If bronchoalveolar lavage or transbronchial biopsy is contemplated, the lung apices demonstrate the largest areas of involvement. (Over) DATE: [**3294-3-23**] 8:47 AM CT CHEST W/CONTRAST; CT 100CC NON IONIC CONTRAST Clip # [**Clip Number (Radiology) 3519**] Reason: evolution of ground glass opacities, ? DAH, ? CHF Admitting Diagnosis: PROSTATE CANCER;PNEUMONIA Contrast: OPTIRAY Amt: 100	**pneumonia** (Positive/Certain/Acute)

In this notebook we demonstrate a basic document level classification of reports with respect to a single finding (fever). We leverage the convenience of Pandas to read our data from a SQLite database and then use Pandas to add our classification as a new column in the dataframe.

Many of the common pyConTextNLP tasks have been wrapped into functions contained in the radnlp pacakge. We important multiple modules that will allow us to write concise code.

```
In [2]:
import pyConTextNLP.pyConTextGraph as pyConText
import pyConTextNLP.itemData as itemData
import pymysql
import numpy as np
import os
import radnlp.io  as rio
import radnlp.view as rview
import radnlp.rules as rules
import radnlp.schema as schema
import radnlp.utils as utils
import radnlp.split as split
import radnlp.classifier as classifier
```

```
import pandas as pd from IPython.display
import clear_output, display, HTML, Image
from IPython.html.widgets
import interact, interactive, fixed
from IPython.display import clear_output
import ipywidgets as widgets
from radnlp.data import classrslts
import networkx as nx
conn = pymysql.connect(host="mys
ql", port=3306,user="jovyan",
passwd='jovyan',db='mimic2')
/opt/conda/lib/python3.5/site-packages/IPython/
html.py:14: ShimWarning: The `IPython.html` package
has been deprecated. You should import from `note-
book` instead. `IPython.html.widgets` has moved to
`ipywidgets`. "`IPython.html.widgets` has moved
to `ipywidgets`.", ShimWarning)
In [3]:
conn = pymysql.connect(host="mys
ql", port=3306,user="jovyan",
passwd='jovyan',db='mimic2')
```

4.7. ANNOTATION COLORING SCHEME

Later in the chapter we are going to provide an HTML display of the marked up documents. Here we specify what colors we want to use for each category we have defined.

```
In [15]:
colors={"critical_finding":"blue",
"pneumonia":"blue", "pneumothorax":"blue",
"diverticulitis":"blue", "definite_ne-
gated_existence":"red", "probable_negat-
ed_existence":"indianred", "ambivalent_
existence":"orange", "probable_existence":"
forestgreen", "definite_existence":"green",
"historical":"goldenrod", "indication":"Pink",
"acute":"golden"}
```

Explanation of getOptions

This is just kind of a port of a command line application where I'd use argparse to get the options.

```
In [37]:
def getOptions():    """Generates arguments for specifying database and other parameters"""    options = {}
options['lexical_kb'] = ["https://raw.githubusercontent.com/chapmanbe/pyConTextNLP/master/KB/lexical_kb_nlm.tsv"]  options["schema"] = "https://raw.githubusercontent.com/chapmanbe/pyConTextNLP/master/KB/schema2.csv"    options["rules"] = "https://raw.githubusercontent.com/chapmanbe/pyConTextNLP/master/KB/classificationRules3.csv"    return options
```

4.8. DEFINE OUR DOMAIN TERMS

You need to define four-tuples for the targets we want to find and the modifiers we are interested in

Each four-tuple needs the following items

- a **lexical term** (e.g., "pulmonary embolism")
- a **category** (e.g., "critical_finding")
- a **regular expression** to capture the lexical term (can be an empty string)
- a **rule** that describes which direction the modifier operates in the sentence (an empty string for a target)

```
# Edit these lines in the code block below
targets.extend([["pulmonary embolism", "CRITICAL_FINDING", "", ""],                    ["pneumonia", "CRITICAL_FINDING", "", ""]]) modifiers.extend((["no definite", "PROBABLE_NEGATED_EXISTENCE", "", "forward"],))
In [38]:
def get_kb_rules_schema(options):
"""    Get the relevant kb, rules, and schema.
"""                    _radnlp_rules    =    rules.read_rules(options["rules"])
_schema = schema.read_schema(options["schema"])
modifiers = itemData.itemData()
```

```
targets = itemData.itemData()
for kb in options['lexical_kb']:
     modifiers.extend( itemData.instantiateFromCSVt
oitemData(kb) )
     targets.extend([["pulmonary embolism", "CRIT-
ICAL_FINDING", "", ""],                    ["pneu-
monia", "CRITICAL_FINDING", "", ""]])    modifiers.
extend((["no   definite",   "PROBABLE_NEGATED_EXIS-
TENCE", "", "forward"],))
     return {"rules":_radnlp_rules,           "sche-
ma":_schema,                   "modifiers":modifiers,
"targets":targets}
In [39]:
def  analyze_report(report,  modifiers,  targets,
rules, schema):
"""     given an individual radiology report, cre-
ates a pyConTextGraph    object that contains the
context markup    report: a text string containing
the radiology reports     """
markup   =   utils.mark_report(split.get_sentences(
report),                                 modifiers,
targets)
clssfy =   classifier.classify_document_targets(mar
kup,                                            rul
es[0],                                          rul
es[1],                                          rul
es[2],                                      schema)
return  classrslts(context_document=markup,  exam_
type="ctpa",  report_text=report,  classification_
result=clssfy)
In [40]:
options = getOptions()
kb = get_kb_rules_schema(options)
#data = data.dropna()
```

4.8.1. Read the Data

We are using the MIMICII Clinic Data, the complete electronic medical record for ICU patients. We are only using the MIMIC2 demo data set as no data use agreement is required, as the records are for 4000 deceased individuals. The data are stored in a MySQL database. We use Pandas to read the reports in, limiting ourselves to radiology reports for individuals with an ICD9 code of pulmonary embolism.

```
In [41]:
data = pd.read_sql("""
SELECT noteevents.subject_id,                    no-
teevents.text,                             icd9.code
FROM noteevents I
NNER JOIN icd9
ON                          noteevents.subject_id =
icd9.subject_id
WHERE (   icd9.code LIKE '415.1%'                 )
AND  noteevents.category  =  'RADIOLOGY_REPORT'""",
conn).drop_duplicates()
data.head(10)
Out[41]:
```

4.8.2. Find Impression Sections and Do Simple Search for Chest and CT

While these radiology reports are all for patients with a pulmonary embolism diagnosis, the images are acquired for lots of reasons. To try to enrich our data set we do some key word filtering to get chest-oriented images.

```
In [42]:
doc_split=["IMPRESSION:", "INTERPRETATION:", "CON-
CLUSION:"]
def find_impression(text, split):
for term in split:
     if term in text:
     return text.split(term)[1]
return np.NaN
data["impression"] = data.apply(lambda row: find_
impression(row["text"], doc_split), axis=1)
data = data.dropna(axis=0, inplace=False)
data["chest"] = data.apply(lambda x: 'chest' in
x["text"].lower() and 'ct' in x["text"].lower(),
axis=1)
```

```
data = data[data["chest"] == True]
data = data.reset_index()
print(data.shape)
data.head()
(661, 6)
Out[42]:
```

4.9. DOCUMENT MARKUP AND CLASSIFICATION

We are going to be doing a lot behind the scenes here. We've got some code that is going to take our lexical and domain definitions, markup our documents according to these definitions and use pyConTextNLP to assign relationships between the concepts for each sentence int he document. Finally, we take a schema that defines how we are going to classify our documents, based on the sentence level markup. The schema we are going to use here is an ordinal schema where we account for existence, uncertainty, and acuity. The document classification is basically a maximum function: the document is assigned the maximum score of any sentence in the document.

We now need to apply our schema to the reports. Since our data is in a Pandas data frame, the easiest way to process our reports is with the DataFrame apply method.

- We use lambda to create an anonymous function which basically just applies analyze_report to the "impression" column with the modifiers, targets, etc. that we have read in separately.
- analyze_report returns a dictionary with keys as any identified targets defined in the "targets" file and values as a tuple with values:

```
In [46]:
data["pe rslt"] = data.apply(lambda x: analyze_repo
rt(x["impression"],                              kb
["modifiers"],                                  kb[
"targets"],                                   kb["ru
```

```
les"],                                    kb["schema"]),
axis=1)
data.head()
Out[46]:

In [44]:
def view_markup(reports, colors):
@interact(i=widgets.IntSlider(min=0,
max=len(reports)-1))
def _view_markup(i):
markup = reports["pe rslt"][i]
rview.markup_to_pydot(markup)
display(Image("tmp.png"))
mt = rview.markup_to_html(markup, color_map=colors)
display(HTML(mt))
In [45]:
view_markup(data, colors)
```

PE Finder Case Review

Report	**Classification**
1) Satisfactory position of lines and tubes as described above. No pneumothorax. 2) Progression of diffuse bilateral air space disease, most prominent in left mid and lower lung zone and right apex. This most likely represents multifocal pneumonia in a patient with fever and infected sputum.	**critical_finding** (Positive/ Certain/Acute)

4.10. RADIOLOGY NLP

4.10.1. Rad (as in cool) NLP

RadNLP is a package that builds upon the pyConTextNLP algorithm's sentence-level text processing to perform simple document-level classification. The package also contains a number of functions for identifying sections of reports, identifying and eliminating boiler-plate text, etc.

In this section I will demonstrate *radnlp's* most basic functionality: *given a text of interest create an overall document classification*. The classification uses a simple maximum function: for each concept marked in a report the maximal schema value occurrence is selected to characterize the report for that concept.

4.10.2. Report Schema and Maximal Value

The document classification is based on schema that combines multiple concepts (e.g., existence, certitude, severity) into a single ordinal scale. The RadNLP GitHub repository includes a knowledge base directory (KBs) contains the schema we have previously developed for critical findings projects. It is included below:

```
# Lines that start with the # symbol are comments
and are ignored
# The schema consists of a numeric value, followed
by a label (e.g., "AMBIVALENT"),
# followed by a Python express that can evaluate to
True or False
# The Python expression uses LABELS from the rules.
processReports.py will substitute
# the LABEL with any matched values identified from
# the corresponding rules 1,AMBIVALENT,DISEASE_
STATE == 2 2,Negative/Certain/Acute,DISEASE_STATE
== 0 and CERTAINTY_STATE == 1 3,Negative/Uncer-
tain/Chronic,DISEASE_STATE == 0 and CERTAINTY_
STATE == 0 and ACUTE_STATE == 0 4,Positive/Uncer-
tain/Chronic,DISEASE_STATE == 1 and CERTAINTY_STATE
== 0 and ACUTE_STATE == 0 5,Positive/Certain/
Chronic,DISEASE_STATE == 1 and CERTAINTY_STATE
== 1 and ACUTE_STATE == 0 6,Negative/Uncertain/
Acute,DISEASE_STATE == 0 and CERTAINTY_STATE == 0
7,Positive/Uncertain/Acute,DISEASE_STATE == 1 and
CERTAINTY_STATE == 0 and ACUTE_STATE == 1 8,Posi-
tive/Certain/Acute,DISEASE_STATE == 1 and CERTAIN-
TY_STATE == 1 and ACUTE_STATE == 1
```

A key idea is "a Python express that can evaluate to True or False."

The radnlp.schema subpackage contains functions for reading schema and applying the schema to the pyConTextNLP findings given rules specified by the user.

There are two key functions in radnlp.schema:

```
def instantiate_schema(values, rule):
"""     evaluates rule by substituting values into
rule and evaluating the resulting literal.    This
is currently insecure           *
"For security the ast.literal_eval() method should
be used."       """
      r = rule
      for k in values.keys():
            r  =  r.replace(k,  values[k].__str__())
#return
      ast.literal_eval(r)
      return eval(r)
def assign_schema(values, rules):
"""
"""
      for k in rules.keys():
            if  instantiate_schema(values,  rules[k]
[1]):
      return k
```

For any given category (e.g., pulmonary_embolism), the maximal schema score encountered in the report is selected to characterize that report.

4.10.3. radnlp RULES

radnlp uses rule files to specify rules that define how particular pyConTextNLP findings relate to radnlp concepts. For example, in the classificationRules3.csv provided in KBs, we provide a rules that state:

The default disease state is 1.

PROBABLE_EXISTENCE AND DEFINITE_EXISTENCE map to a disease state of 1

Rules as currently indicated are not quite general and reflect particular use cases we were working on.

4.10.4. Types of Rules

- CLASSIFICAITON_RULE: these are the rules that relate to disease, temporality, and certainty
- CATEGORY_RULE: these are only partially developed concepts that attempt to address combinatorics problems in pyConTextNLP

by making default findings more general (e.g., infaract) and then tries to create more specific findings by attaching anatomic locations to the findings (e.g., an infarct becomes a critical finding when attached to an anatomic concept like brain_anatomy or heart_anatomy.

- SEVERITY_RULE: Again, not fully developed but relates to extracting severity concepts (e.g., large or 4.3 cm).

```
# Lines that start with the
# symbol are comments and are ignored,,,,,,,,,,,,,
# processReport current has three types of rules: @
CLASSIFICATION_RULE, @CATEGORY_RULE, and @SEVERITY_
RULE,,,,,,,,,,,, # classification rules would be for
things like disease_state, certainty_state, tempo-
rality state,,,,,,,,,,,, # For each classification_rule
set," there is a rule label (e.g., ""DISEASE_STATE"."
This must match",,,,,,,,,,,,, # the terms used in the
schema file,,,,,,,,,,,,, # Each rule set requires a
DEFAULT which is the schema value to be returned
if no rule conditions are satisifed,,,,,,,,,,,,, #
Each rule set has zero or more rules consisting of
a schema value to be returned if the rule evaluates
to true,,,,,,,,,,,,, # A rule evalutes to true if
the target is modified by one or more of the ConText
CATEGORIES listed following,,,,,,,,,,,,, @CLASSIFI-
CATION_RULE,DISEASE_STATE,RULE,0,DEFINITE_NEGATED_
EXISTENCE,PROBABLE_NEGATED_EXISTENCE,FUTURE,INDIC
ATION,PSEUDONEG,,,,, @CLASSIFICATION_RULE,DISEASE_
STATE,RULE,2,AMBIVALENT_EXISTENCE,,,,,,,,, @CLAS-
SIFICATION_RULE,DISEASE_STATE,RULE,1,PROBABLE_
EXISTENCE,DEFINITE_EXISTENCE,,,,,,,, @
CLASSIFICATION_RULE,DISEASE_STATE,DEFAULT,1,,,,,,,,,,@CLAS-
SIFICATION_RULE,CERTAINTY_STATE,RULE,0,PROBABLE_
NEGATED_EXISTENCE,AMBIVALENT_
EXISTENCE,PROBABLE_EXISTENCE,,,,,,, @
CLASSIFICATION_RULE,CERTAINTY_STATE,DEFAULT,1,,,,,,,,,,
@CLASSIFICATION_RULE,ACUTE_STATE,RULE,0,HIS
TORICAL,,,,,,,,, @CLASSIFICATION_RULE,ACUTE_
STATE,DEFAULT,1,,,,,,,,,,, #CATEGORY_RULE rules
specify what Findings (e.g., DVT) can have the
```

```
category modified by the following ANATOMIC modi-
fies,,,,,,,,,,,,, @CATEGORY_RULE,DVT,LOWER_DEEP_
VEIN,UPPER_DEEP_VEIN,HEPATIC_VEIN,PORTAL_SYS-
TEM_VEIN,PULMONARY_VEIN,RENAL_VEIN,SINUS_
VEIN,LOWER_SUPERFICIAL_VEIN,UPPER_SUPERFI-
CIAL_VEIN,VARICOCELE,ARTERIAL,NON_VASCULAR @
CATEGORY_RULE,INFARCT,BRAIN_ANATOMY,HEART_
ANATOMY,OTHER_CRITICAL_ANATOMY,,,,,,,,, @CATEGORY_
RULE,ANEURYSM,AORTIC_ANATOMY,,,,,,,,,,, #SEVERITY_
RUlE specify which targets to try to obtain severity
measures for,,,,,,,,,,,,, @SEVERITY_RULE,AORTIC_
ANATOMY_ANEURYSM,SEVERITY,,,,,,,,,,,
In [ ]:
%matplotlib inline
In [ ]:
import radnlp.rules as rules
import radnlp.schema as schema
import radnlp.utils as utils
import radnlp.classifier as classifier
import radnlp.split as split
from IPython.display import clear_output, display,
HTML
from IPython.html.widgets import interact, interac-
tive, fixed
import io from IPython.html
import widgets # Widget definitions
import pyConTextNLP.itemData as itemData
from pyConTextNLP.display.html import mark_docu-
ment_with_html
Below are two example radiology reports pulled from
the MIMIC2 demo data set.
In [ ]:
reports = ["""1. Pulmonary embolism with filling
defects noted within the upper and lower lobar
branches of the right main pulmonary artery. 2.
Bilateral pleural effusions, greater on the left.
3. Ascites. 4. There is edema of the gallblad-
der wall, without any evidence of distention,
```

```
intra- or extra-hepatic biliary dilatation. This, along with    stranding within the mesentery, likely represents third spacing of fluid.     5. There are several wedge shaped areas of decreased perfusion within the     spleen, which may represent splenic infarcts.          Results were discussed with Dr. [**First Name8 (NamePattern2) 15561**] [**Last Name (NamePattern1) 13459**]      at 8 pm on [**3099-11-6**].""",               """1. Filling defects within the subsegmental arteries in the region     of the left lower lobe and lingula and within the right lower lobe consistent      with pulmonary emboli. 2. Small bilateral pleural effusions with associated bibasilar atelectasis.                3. Left anterior pneumothorax.            4. No change in the size of the thoracoabdominal aortic aneurysm.             5. Endotracheal tube 1.8 cm above the carina. NG tube within the stomach,     although the tip is pointed superiorly toward the fundus.""",            """1. There are no pulmonary emboli observed.            2. Small bilateral pleural effusions with associated bibasilar atelectasis.                3. Left anterior pneumothorax.            4. No change in the size of the thoracoabdominal aortic aneurysm.             5. Endotracheal tube 1.8 cm above the carina. NG tube within the stomach,     although the tip is pointed superiorly toward the fundus.""" ]
In [ ]:
#!python -m textblob.download_corpora
```

4.10.5. Define Locations of Knowledge, Schema, and Rules Files

```
In [ ]:
def getOptions():   """Generates arguments for specifying database and other parameters"""   options = {}   options['lexical_kb'] = ["https://raw.githubusercontent.com/chapmanbe/pyConTextNLP/master/KB/lexical_kb_04292013.tsv",              "https://raw.githubusercontent.com/chapmanbe/pyConTextNLP/
```

```
master/KB/criticalfinder_generalized_modifiers.tsv"]
options['domain_kb'] = ["https://raw.githubuser-
content.com/chapmanbe/pyConTextNLP/master/KB/
pe_kb.tsv"]#[os.path.join(DATADIR2,"pe_kb.tsv")]
options["schema"] = "https://raw.githubusercontent.
com/chapmanbe/RadNLP/master/KBs/schema2.csv"#"file
specifying schema"    options["rules"] = "https://
raw.githubusercontent.com/chapmanbe/RadNLP/master/
KBs/classificationRules3.csv" # "file specifying sen-
tence level rules")    return options
```

4.10.6. Define Report Analysis

For every report we do two steps

- Markup all the sentences in the report based on the provided targets and modifiers
- Given this markup we apply our rules and schema to generate a document classification.

radnlp provides functions to do both of these steps:

- radnlp.utils.mark_report takes lists of modifiers and targets and generates a pyConTextNLP document graph
- radnlp.classify.classify_document_targets takes the document graph, rules, and schema and generates document classification for each identified concept.

Because pyConTextNLP operates on sentences we split the report into sentences. In this function we use radnlp.split.get_sentences which is simply a wrapper around textblob for splitting the sentences.

```
In [ ]:
def  analyze_report(report,  modifiers,  targets,
rules, schema):
"""    given an individual radiology report, cre-
ates a pyConTextGraph    object that contains the
context markup    report: a text string containing
the radiology reports
"""

markup  =  utils.mark_report(split.get_sentences(
report),                                modifiers,
targets)
return
```

```
classifier.classify_document_targets(markup,
rules[0],                                              ru
les[1],                                               rul
es[2],                                            schema)
In [ ]:
def process_report(report):
     options = getOptions()
     _radnlp_rules           =          rules.read_
rules(options["rules"])
     _schema=schema.read_schema(options["schema"])
     #_schema = readSchema(options["schema"])
     modifiers = itemData.itemData()
     targets = itemData.itemData()
     for kb in options['lexical_kb']:
          modifiers.extend( itemData.instantiateFro
mCSVtoitemData(kb) )
     for kb in options['domain_kb']:
          targets.extend( itemData.instantiateFrom
CSVtoitemData(kb) )
     return analyze_report(report, modifiers, tar-
gets, _radnlp_rules, _schema)
In [ ]:
rslt_0 = process_report(reports[0])
```

radnlp.classifier.classify_document_targets returns a dictionary with keys equal to the target category (e.g., pulmonary_embolism) and the values a 3-tuple with the following values:

- The schema category (e.g., 8 or 2).
- The XML representation of the maximal schema node
- A list (usually empty (not really implemented yet)) of severity values.

```
In [ ]:
for key, value in rslt_0.items():
     print(("%s"%key).center(42,"-"))
     for v in value:
          print(v)
In [ ]:
rslt_1 = main(reports[1])
for key, value in rslt_1.items():
```

```
    print(("%s"%key).center(42,"-"))
    for v in value:
        print(v)
```

4.10.7. Negative Report

For the third report I simply rewrote one of the findings to be negative for PE. We now see a change in the schema classification.

```
In [ ]:
rslt_2 = main(reports[2])
for key, value in rslt_2.items():
    print(("%s"%key).center(42,"-"))
    for v in value:
        print(v)
In [ ]:
keys = list(pec.markups.keys())
keys.sort()
pec.reports.insert(pec.reports.columns.get_loc(u'
markup')+1,                    "ConText Coding",
[codingKey.get(pec.markups[k][1].get("pulmonary_
embolism",[None])[0],"NA") for k in keys])
```

4.11. CONCLUSIONS

In this chapter we have focused on NLP applied to Biological contexts, in particular healthcare and biomedical scenarios. We have provided an example-based approach, which should help readers with a deep interest in computation.

5 CHAPTER

DEEP LEARNING

CONTENTS

We have introduced NLP in previous chapter. We now introduce to the readers what is considered as the next big thing in the area of computational biology – *Deep Learning*. This chapter is aimed at introducing the core concepts and would serve as 101 materials. The reader is encouraged to delve deep into the matter and expand the wings to other areas of biomedical informatics using Deep Learning.

So let's get started with a brief introduction of Deep Learning.

5.1. WHAT IS DEEP LEARNING?

Over the past 5 years, the catch phrase *Deep learning*, has burgeoned into a very big universe. It is no longer considered a somewhat 'niche' field, as it was a few years back, comprising of a cloistered group of researchers. The traditional field of Machine Learning (ML) is being replaced swiftly by Deep Learning (DL). A large variety of tasks have been accomplished by DL, and many a times they are difficult to comprehend and explain.

Interest in DL has sky-rocketed, with constant coverage in the popular media. Researchers employ DL routinely. And papers in top journals like Science, Nature, Nature Methods and JAMA feature DL on a regular basis.

For those interested in understanding the current developments in DL, an attendance at the Annual Conference on Neural Information Processing Systems (NIPS) is recommended. NIPS is the main conference for deep learning research and has historically been where a lot of the new methodological research gets published. Interest in the conference has surged during the 5 years:

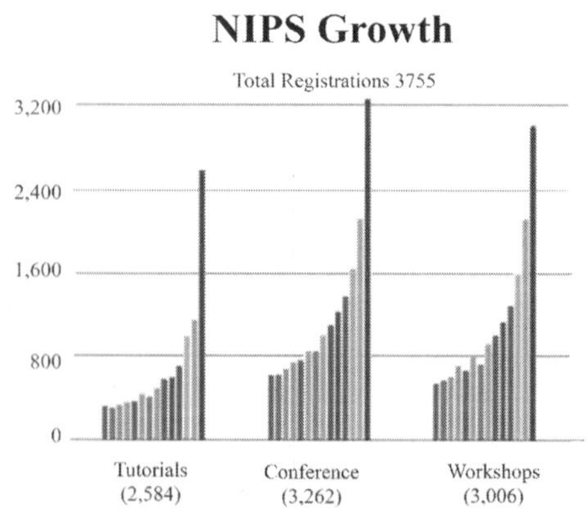

The above chart only goes to 2015 and this year's NIPS had over 6,000 attendees. This trend is a clear sign of increasing or rather surging interest in DL. This is a clear confirmation that we were entering a period of serious hype occurred during NIPS 2013 when Mark Zuckerberg made a surprise visit to recruit deep learning talent, and ended up convincing Yann Lecun to be the director of the Facebook AI lab. Now, let us see if we can understand all this in bits and pieces. Let us ask some important questions.

Does this hype mean anything?

Does it happening now mean anything?

Deep Learning is a result of the constant toil and sweat of dedicated scientists in the world. It has been quite successful and certainly it is the next big thing. Although the main ideas behind deep learning have been in place for decades, it wasn't until data sets became large enough and computers got fast enough that their true power could be revealed. I think it is time to take a quick look at the historical events that led to the events that unfolded DL to the world at large.

5.1.1. Milestones in the Development of Neural Networks

We show below a pictorial time line of the major milestones from Neural Networks to DL.

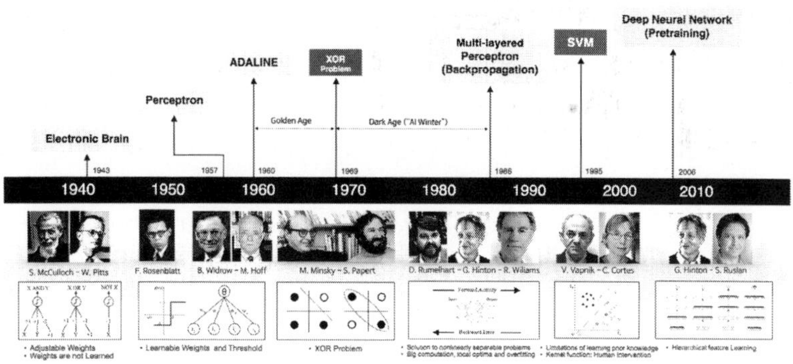

5.1.2. In The Beginning (1940s)

Now, let's see what really happened. The idea to create a thinking machine started almost at the same time, as was the discovery of modern computing machine. The 'Turing Test' as proposed by Alan Turing in his seminal paper 'Computing Machinery and Intelligence' lays down some criteria for the assessment of a machine as intelligent. Machine Learning, in initial years was largely influenced by the then working theories of the brain Walter Pitts and Warren McCulloch were the first to take up the challenge in ML. A concept known as "thresholded logic unit" was designed by them to mimic the way a neuron was thought to work. This was a theme which was used in many areas after their proposal.

Frank Rosenblatt's "perceptron" was the first real precursor to modern neural networks unequivocally. Certainly, at the point of time when it was unveiled it was pretty impressive. Briefly, the perceptron came with a learning procedure that would converge to the correct solution and could recognize letters and numbers. Rosenblatt was so confident that the perceptron would lead to true AI, that in 1959 he remarked:

[The perceptron is] the embryo of an electronic computer that [the Navy] expects will be able to walk, talk, see, write, reproduce itself and be conscious of its existence.

5.1.3. The First AI Winter (1969)

By 1950's and 1960's, since the proclamation of Rosenblatt's perceptron, it had acquired quite a following and had garnered quite a bit of attention. Marvin Minsky, often regarded as father of AI, sensed that something was off with perceptron. To quote Minsky:

However, I started to worry about what such a machine could not do. For example, it could tell 'E's from 'F's, and '5's from '6's—things like that. But when there were disturbing stimuli near these figures that weren't correlated with them the recognition was destroyed.

Along with the double-PhD wielding Seymor Papert, Minksy wrote a book entitled *Perceptrons* that effectively killed the perceptron. This conclusively ended the embryonic idea of a neural net. It took them to shut it down by just showing that perceptron was incapable of learning a simple exclusive-or (XOR) function. Even worse was the fact that they *proved* that it was theoretically impossible for it to learn such a function, no matter how long you let it train. Today, we know that XOR is a non-linear function. And it isn't surprising to us, as the model implied by the perceptron is a linear one. But, at the time this was enough to kill all research on neural nets and usher in the first AI winter. So, their take on AI was actually detrimental.

5.1.4. The Backpropagandists Emerge (1986)

Good thoughts and constructive ideas never fade away. The lure of perceptron was there amongst the scientific community. It was not until 1980's that the Minsky-induced snowpack began to melt. The signs of life began to return to the neural net community. George Bool's great-great-grandson, a man by the name of Geoff Hinton, finished his PhD studying neural networks in 1978. And by 1986 had moved to the Institute for Cognitive Science at UC San Diego. By that time Neural Nets had gained popularity. A seminal contribution was made by Hinton, along with David Rumelhart and Ronald Williams, Hinton published a paper entitled "Learning representations by back-propagating errors." In this paper they showed that neural nets with many hidden layers could be effectively trained by a relatively simple procedure. This would allow neural nets to get around the weakness of the perceptron; because the additional layers endowed the network with the ability to learn nonlinear functions. Almost at the same time it was revealed that there is something called **universal approximation theorem**. This had the ability to learn *any* function. Probably, this was the turning point in the field of neural nets.

Lets understand the algorithm in a simple language. Basically, the algorithm works by taking the derivative of the network's loss function and back-propagating the errors and uses it to update the parameters in the lower layers, hence the common name **backprop**. This algorithm lead to some early successes. Particularly, cases involving training **convoluational neural nets** (CNNs, also referred to using the less pronounceable convents) to recognize handwritten digits was extremely successful. But, even with all this neural nets headed for a second ice age. *The approach didn't scale to larger problems*. By the 90s the support vector machine (SVM) became the method of choice, and neural nets were moved back into storage. In hindsight, neural nets were simply an idea before their time, but it would be another 10-15 years before data and computing power were such that neural nets would finally be able to reach their final form.

5.1.5. Rebranding as 'Deep Learning' (2006)

The ice age in the area of Neural nets received a bit of sunshine around 2006. This time again Hinton championed the cause by declaring that he knew of the functioning of brain, and introduced the ideas of *unsupervised pretraining* and *deep belief nets*. Simply put, it involved

- training a simple 2-layer unsupervised model like a restricted Boltzmann machine;
- freeze all the parameters;

- stick on a new layer on top;
- train *just the parameters* for the new layer;
- keep adding and training layers in this greedy fashion until you have a deep network;
- then use the result of this process to initialize the parameters of a traditional neural network.

Using this strategy, people were able to train networks that were deeper than previous attempts, prompting a rebranding of **'neural networks' to 'deep learning.'** If you go to deeplearning.net, which I believe is owned by MILA, the title proudly declares

Deep Learning… moving beyond shallow machine learning since 2006!

in recognition of this transition.

5.1.6. The Breakthrough

The concepts of unsupervised pretraining and deep belief nets resulted in a spurge of papers from several research labs. The real breakthrough was yet to come. It appeared in an area where deep nets were employed for speech recognition, and a neural model was constructed. There is a saying:

Neural networks are the second best way to do almost anything

With all things going good for Neural Networks, there came a halt. Things would finally come to a head in 2012 on the Large Scale Visual Recognition Challenge (LSVRC). In 2010, a large database known as **Imagenet** containing millions of labeled images was created and published by Fei-Fei Li's group at Stanford. This database was coupled with the annual LSVRC, where contestants would build computer vision models, submit their predictions, and receive a score based on how accurate they were. In the first two years of the contest, the top models had error rates of 28% and 26%. In 2012, Alex Krizhevsky, Ilya Sutskever, and Geoff Hinton entered a submission that would halve the existing error rate to 16%. The model combined several critical components that would go one to become mainstays in deep learning models.

The main attraction of the model was the usage of graphics processing units (GPUs) to train the model.

What are GPUs?

Let us answer this in parts.

1. GPUs are essentially parallel floating-point calculators with 100s–1000s of cores;
2. The speed-up offered by GPUs meant they could train larger models, which led to lower error rates;
3. They also introduced a method to reduce over fitting known as dropout and;
4. They have also employed the rectified linear activation unit (ReLU),

both of which are now bread and butter components in modern deep learning. The network went on to become known as "Alexnet" and the paper describing it has been cited over 10,000 times since it was published at NIPS in 2012. Here is a graph showing the error rate on Imagenet over time, note the precipitous drop in 2012 and the huge uptick in teams using GPUs:

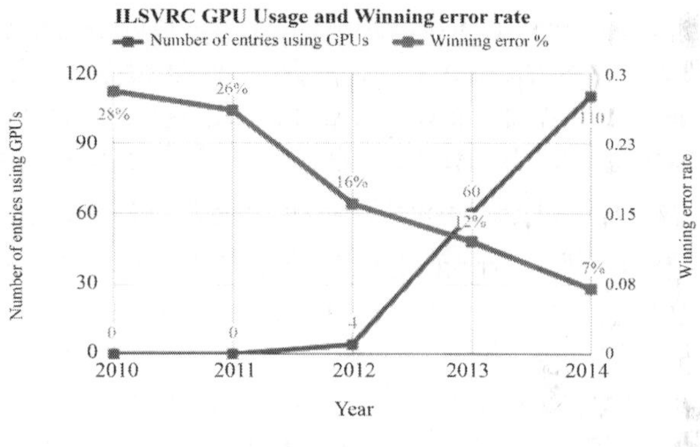

Many important innovations have appeared after this result, but I think it's safe to say this is the moment when the deep learning levy finally broke. Many are becoming convinced that we are in the midst of a Great AI Awakening and billions are being spent by companies to stockpile AI talent and technology. It's hard to know what will come next and the extent to which deep learning will play a role, but for now, it does feel like we've experience a paradigm shift in machine learning that is here to stay.

5.1.7. Why the Craze?

By now, it is clear to you that all the theoretical concepts were in place in 80s and 90s. So what really happened during the past few years? It need not be over emphasized that although there are many factors, the two most crucial components are the availability of **massive labeled data sets** and

GPU computing. Given below is a list of factors with a role in the deep learning revolution:

1. **Appearance of large, high-quality labeled datasets** – Data serves as a fuel. And with the current technological advances, handling large data being generated at a regular pace is not difficult.
2. **Massively parallel computing with GPUs** – At the core of neural nets is a bunch of floating point calculations. Which when executed in parallel drastically reduce the time to execution. GPUs have emerged as the best architectures for performing such calculations. This transition from CPU-based to GPU-based computations has resulted in massive speed-ups for computational models.
3. **Activation functions which are backdrop-friendly** – The use of ReLU have reduced the vanishing gradient problem, effectively replacing activation functions like tanh and logit.
4. **Software platforms** – With the advent of constantly improving frameworks like Tensorflow, Theano, Chainer, and Mxnet have provided an integrated platform for GPU computing and making prototyping error prone. Essentially, they are designed to take care of gradients and GPU management. Helping you to focus on the model.
5. **New regularization techniques** – There is constant description and implementation of new techniques which allow training of larger networks without overfitting.
6. **Robust optimizers** – There is a lot of emphasis on designing robust optimizers.

One should bear in mind that most of these improvements have been driven by empirical performance on a standard set of benchmarks. Hence, there is no theoretical justification to any of these techniques.

The success of deep learning is largely a success of engineering. I think this statement is not an overstatement. I would like to add computer science and programming to this as well. Certainly, the presence of extremely brilliant programmers is the reason for the current state-of-the-art implementation of software platforms.

As expected, the field moves fast. Any research gets old on a timescale of *months*, hence to keep your finger on the pulse can be tough. Here are a few suggestions which one can use to stay current with the developments:

1. A better place to start is reading articles being published in ***arXiv***. Specifically, focus on subsection on machine learning. Almost all the papers on deep learning get posted on arXiv. Hence, cutting edge discoveries can be checked quickly. We leave it to the imagination of the reader as to how to assimilate the knowledge being generated at such a fast pace.
2. Andrej Karpathy has developed a tool called *arxiv-sanity* to help you shift through the vast volume of research. It makes it easy to find papers relevant to your interests. Even gives a Netflix style recommendation.
3. The *keras* blog details on the new implements of the new developments in *keras*.
4. Google research blog – This is a great place discussing all of the research going on in Google.
5. Highly recommended are podcasts like *talking machines*.
6. **Twitter** – this is probably the **best** way to stay up to date, because everything happens in real time. Here is an incomplete list of who to follow to help get you started:

a. Yann Lecun – Head of Facebook's AI research lab. Giant of the field and tweets often. If you're looking for some laughs or just want to #feelthelearn, check out the parody account Bored Yann Lecunn run by Jack Clark

b. Andrej Karpathy – Research scientist at OpenAI, also has a great blog

c. Ian Goodfellow – Researcher at OpenAI, creator of generative adversarial networks(i.e. GANs), explainer of adversarial examples, world's most trustworthy last name

d. Francois Chollet – Creator of keras, researcher at Google

e. Russ Salakhutdinov – Director of AI Research at Apple and professor at CMU

f. Fei-Fei Li – Imagenet creator, Stanford professor and director of the Stanford AI lab, Chief Scientist at Google Cloud

g. Andrew Ng – Cofounder of coursera, former Standford professor, Chief scientist at Baidu

h. Sander Dieleman – Creator of the lasagne deep learning framework and research scientist at Deep Mind. Also has a great blog with some epic Kaggle contest write ups. You can learn a lot just by reading how Sander approaches a Kaggle contest.

i. Hugo Larochelle – Google Brain researcher.

5.2. INSIDE DEEP LEARNING

Ivakhnenko and Lapa in 1965 were the first to employ the earliest deep-learning-like algorithms that had multiple layers of non-linear features (Figure 5.1). They have used thin but deep models with polynomial activation functions which they had analyzed with statistical methods. They had employed statistical methods to select best features in each layer and forward them to the next layer. But, they did not use backpropagation to train their network end-to-end. Instead, they have used layer-by-layer least squares fitting where previous layers were independently fitted from later layers.

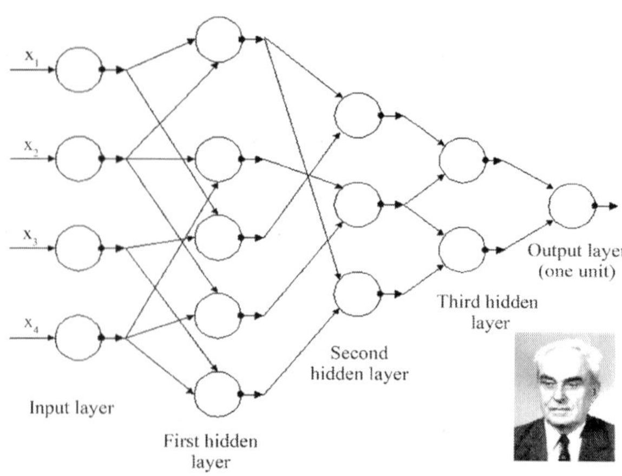

Figure 5.1: One of the first pictures of first generation Deep Networks.

Digging into literature reveals that one of the first known deep network architecture was trained by Alexey Grigorevich Ivakhnenko in 1965. In this particular architecture, the feature selection steps after every layer leads to an ever-narrowing architecture. This terminates when no further improvement can be achieved by the addition of successive layers. Essentially, there is an emphasis on convergence.

Now, let us come to the CNNs. Some of the earliest convolutional networks were used by Fukushima in early 80's. Fukushima can be regarded as one of the first scientists to employ CNNs similar to modern deep networks. His networks had multiple convolutional and pooling layers. This concept of convolutional and pooling layers is very similar to modern neural networks. In this case the network was trained using a reinforcement scheme, which involves a trail of strong activation in multiple layers. Additionally, one would assign important features of each image by hand by increasing the weight on certain connections. It was quite an important advancement.

The concept of error backpropagation to train deep models was still not implemented. An inefficient and incomplete form of backpropagation was derived in early 1960s. Linnainmaa is credited with deriving the first modern form in his 1970 masters thesis. This included FORTRAN code for backpropagation, but it did not delve into its applications to neural networks. Even beyond this point, the concept of backpropagation was relatively unknown. In 1980s as well very few documented applications of backpropagation existed. In 1980's it was shown that in neural networks backpropagation could yield interesting distributed representations (connectionism). This addressed an important question in coginitive psychology.

Whether human cognition relied on connectionism or symbolic logic (computationalism)?

The scientists at Bell Labs were not behind. Lecun showed the first real application of backpropagation. He was the first to use CNNs with backpropagation to classify handwritten digits (MNIST). This was a pioneering advancement and this system was later employed to read large numbers of handwritten checks in the United States.

Any advancement in science and technology needs generous funding. This was the case with the field of computational neural networks as well. The term artificial intelligence dropped to near pseudoscience status during the AI winter and the field still needed some time to recover. Some important advances were made in this time, for example, the long short-term memory (LSTM) for recurrent neural networks by Hochreiter and Schmidhuber in 1997. These advances went mostly unnoticed, being overshadowed by

Support Vector Machine methodologies.

We had to wait till the computers got faster. A major advance was the introduction of graphics processing units (GPUs). This wait was fruitful. Now, the roles reversed and SVMs started facing tough competition with NNs. In a nutshell, Neural networks might be slower when compared to support vector machines but they reach much better results with the same amount of data. There is a bottleneck, neural networks continue to improve with more training data which is quite distinctive.

If everything looks cosy, then probably we have solution for every problem. But, hurdles are many. To quite the main hurdle in in training big and deep networks is the vanishing gradient problem. This book is not attempting to present a quantitative picture of this problem. Rather, we would take a layman's approach to these concepts. Basically, this translates to minimal to no learning in earlier layers in NNs as no signal reached these layers.

Major efforst were directed towards finding a solution to this problem. One of the first solution to this problem was *layer-by-layer pretraining*. In this method a model is built in a layer-by-layer fashion initially. One employs unsupervised learning so that the features in early layers are already initialized or "pretrained" with some suitable features (weights). This training gives a good handle on adjusting features in earlier layers. This leads to good results. Initial attempts in 1990's were by Schmidhuber on approaches concerning pretraining of recurrent neural networks. LSTM was an alternative solution to this problem.

We would stop on historical developments at this point. The literature is vast and it is impossible to cover it in this book with a specific focus on biomedical informatics. But, we would like to say a few words on Perceptron and AlexNet.

5.2.1. Perceptron

In today's neural netwoek community, by perceptron one understands a system comprising of only a single linear or nonlinear unit. If one takes a perceptron with a nonlinear unit, which has been trained with the delta rule then it can find the nonlinear plane that separates data points of two different classes. This is a geometrical picture of perceptron. Even if such a plane is missing, the perceptron will often produce results provide good classification accuracy with some planes. This success came to a halt, when it was proven that the perceptron fails miserably to separate seemingly simple patterns like the ones provided by XOR function.

5.2.2. AlexNet

AlexNet is a convolutional network architecture named after Alex Krizhevsky, who along with Ilya Sutskever under the supervision of Geoffrey Hinton applied this architecture to the ILSVRC-2012 competition that featured the ImageNet dataset. They improved the convolutional network architecture developed by Ciresan and colleagues, which won multiple international competitions in 2011 and 2012 by using *rectified linear units* (ReLU) for enhanced *speed* and *dropout* for improved generalization. Their results stood in stark contrast to feature engineering methods, which immediately created a great rift between deep learning and feature engineering methods for computer vision. AlexNet heralded the mainstream usage and the hype of deep learning, in its second improved appearance.

We discuss the core concepts of deep learning in the following section.

5.3. CORE CONCEPTS

5.3.1. Machine Learning

In machine learning we usually have a simple protocol, consisting of the following steps:

(1) one starts with some initial data,

(2) then one train's a model on that data, and

(3) finally one employs the trained model to make predictions on any new data.

The obious question in your mind now is this:

What is training?

We give a simplistic definition of training as a process in which a model is visualized to be learning as and when it is made to see new and unfamiliar data in a *stepwise* manner. The model then starts making predictions at each individual step, and gets a feedback on the accuracy of predictions. The way this feedback is codified is in the form an *error*. One usually employs some measure like its distance (separation) from the correct solution. This is then employed to correct the errors which are made in predictions.

In the parameter space, the learning process is seen as a game of playing with different parameter sets. *Tweaking* of parameters, as it is known in the ML community, can at times make a seemingly prediction appear wrong and a different choice of parameters can reverse the argument. Hence, it usually takes many iterations to train a model with good predictive performance. The whole objective is to improve the accuracy of predictions. The question is forhow long should this continue. Usually, this iterative predict-and-adjust process continues until the predictions of the model no longer improve. In other words, the process continues until convergence.

5.3.2. Feature Engineering

Data, in most instances, will have some form of patterns inherent in it. Extracting features from data is known as feature engineering. This can also lead to identification of distinct classes while building models using ML. To give a simple example, you would like to consider the number of purplish vs. yellowish pixels. This can be an indicator of whether a sky represents dawn or dusk. This is an important aspect to limit the number of classes while classifying a dataset.

For prediction related topics, feature engineering is an important skill to master. In fact, it is the most important one. However, it is difficult to learn and master. Different datasets require different approaches, hence feature engineering requires to be carried out with utmost care to detail. The features which are found to be good for a particular data set may not be usable for a similar data set. This difficulty of feature engineering has spawned the efforts to find algorithms to automatically engineer features.

5.3.3. Feature Learning

As discussed above, feature engineering involves feature-learning algorithms to detect common patterns in a given data set which distinguish between different classes. It also involves extracting them automatically to be used in a classification or regression process. Essentially, feature learning is looked upon as feature engineering which is carried out automatically by algorithms. Now, let us see how this would be useful in deep learning. As discussed earlier, layers known as convolutional layers are exceptionally good at finding good features in data sets (images) to the next layer to form a hierarchy of nonlinear features that grow in complexity. It is the final layer which uses all these generated features for classification or regression (the last layer in a convolutional net is, essentially, *multinomial logistic regression*).

5.3.4. Deep Learning

Let us see how hierarchical Feature Learning helps in making predictions. We start by extracting multiple layers of non-linear features which are then passed to a classifier which combines all the features to make predictions. In DL, we are interested in stacking very deep hierarchies of non-linear features. We cannot learn complex features from a few layers. To generate features that contain more information we cannot operate on the inputs directly, but we need to transform our first features again to get more complex features that contain more information to distinguish between classes.

Human brain does exactly the same thing:

(a) Visual cortex of brain contains the first hierarchy of neurons. These neirons are sensitive to specific edges and blobs;

(b) Down the visual cortex the brain regions are even more sensitive to complex structures.

Using such hierarchical feature learning, in computational image detection, was present even before the advent of DL. As pointed out earlier vanishing gradient problem is a persistent one in DL. Essentially, it is a situation where gradients became too small to provide a learning signal for very deep layers. Thus, making many of these architectures perform poorly.

The concept of deep learning arose when people started addressing strategies which could be designed to generate deep hierarchies of non-linear features, by overcoming the problems with vanishing gradients, so that one can train architectures with dozens of layers of non-linear hierarchical features. In the early 2010s, it was shown that combining GPUs with activation functions was sufficient to train deep architectures without major difficulties. From this point forward the interest in deep learning grew steadily.

One may now think that DL is just learning deep non-linear hierarchical features. But, you would be surprised to hear that it is not so. It also involves learning to detect very long non-linear time dependencies in sequential data. Major algorithms which deal with sequential data have a memory of the last 10 time steps, long short-term memory (LSTM) recurrent neural networks allow the network to pick up on activity hundreds of time-steps in the past to make accurate predictions. The concept of LSTM was invented by Sepp

Hochreiter and Jürgen Schmidhuber in 1997. Their popularity have sky rocketed since 2013 and together with convolutional nets they form one of two major success stories of deep learning.

Let us now venture into some simple concepts in DL. The first among these is Logistic Regression.

5.3.5. Logistic Regression

Everybody understands that a regression analysis estimates the relationship between statistical input variables in order to predict an outcome variable. Logistic regression (LR) is a regression model that uses input variables to predict a *categorical* outcome variable that can take on one of a limited set of class values.

This type of regression involves application of logistic sigmoid function to weighted input values. This leads to the generation of a prediction of which of two classes the input data belongs to (or in case of multinomial logistic regression, which of the multiple classes).

In the language of DL, LR is quite similar to a non-linear perceptron. Or it can be thought of as a neural network without hidden layers. LR is easy to interpret and also reliable if some of the statistical properties for the input variables are found to hold good. If these statistical properties hold, one can produce a very reliable model with very little input data. This makes logistic regression valuable for areas where *data are scarce*, like the **medical** and **social sciences** where logistic regression is used to analyze and interpret results from experiments.

In deep learning, the final layer of a neural network used for classification can often be interpreted as a logistic regression. In this context, one can see a deep learning algorithm as multiple feature learning stages, which then pass their features into a logistic regression that classifies an input. Hence, this is quite a valuable concept in DL.

5.3.6. Artificial Neural Network (ANN)

This is quite a popular term in the area of computational neuro biology. Let us see what it mean in a simple language. An artificial neural network does the following:

(1) takes some input data, and

(2) transforms this input data by calculating a weighted sum over the inputs and

(3) applies a non-linear function to this transformation to calculate an intermediate state.

The three steps above constitute what is known as a layer, and the transformative function is often referred to as a unit. The intermediate states—often termed *features*—are used as the input into another layer. Through repetition of these steps, the artificial neural network learns multiple layers of non-linear features, which it then combines in a final layer to create a prediction. The neural network learns by generating an error signal that measures the difference between the predictions of the network and the desired values and then using this error signal to change the weights (or parameters) so that predictions get more accurate.

5.3.7. Unit

A layer consists of a unit which is nothing but an activation function. This plays a role of trnaforming the inputs using a nonlinear activation function (for example by the logistic sigmoid function). Usually, a particular unit will have many incoming connections and several outgoing connections. With the introduction of LSTMs the structure of units became even more complex. In this case, a unit can have multiple activation functions. These would then compute the final output over an array of nonlinearly transformed input values. Please bear in mind that other functions in DL like pooling and/or convolution are not referred to as units.

5.3.8. Artificial Neuron

Neuron or more accurately artificial neuron is an equivalent term to *unit*, but implies a close connection to neurobiology and the human brain. Deep learning has very little to do with the brain. The term neuron was popularized to differentiate the neural network (NN) from perceptron. With the media championing the use of word 'neauron' since the successes of deep learning, there is a danger of losing focus and the actual definition of the field. In this case, it is not restricted to human brain, as it is usually explained by media. Hence, the use of the term neuron is discouraged and the more descriptive term *unit* is being encouraged by practitioners.

5.3.9. Activation Function

Now, let us understand what an activation function (AF) is and what role it plays in DL. In simple terms, an activation function (AF) takes in weighted data and outputs a non-linear transformation of the data.

output=median(0,weighted$_{data}$)

This is a rectified linear activation function. Now, a confusion arises as to what is the difference between *units* and *activation functions*. Simply *put a unit can have multiple activation functions*. The difference between linear and non-linear activation functions is shown below with the relationship of some weighted values:

Imagine the four points *X1*, *X2*, *Y1* and *Y2*. The pairs *X1* / *X2*, and *Y1* / *Y2* lie close to each other, but *X1* is distant from *Y1* and *Y2*, and vice versa; the same for *X2*.

With a linear transformation the relationship between pairs might change. For example, *X1* and *X2* might be far apart, but this implies that *Y1* and *Y2* are also far apart. The distance between the pairs might shrink, but if it does, then both *Y1* and *Y2* will be close to *X1* and *X2* at the same time. We can apply many linear transformations, but the relationship between *X1* / *X2* and *Y1* / *Y2* will always be similar.

In contrast, with a non-linear activation function we can increase the distance between *X1* and *X2* while we *decrease* the distance between *Y1* and *Y2*. We can make *Y1* close to *X1*, but *Y2* distant from *X1*. By applying non-linear functions, we create new relationships between the points. With every new non-linear transformation we can increase the complexity of the relationships. In deep learning, using non-linear activation functions creates increasingly complex features with every layer. In contrast, the features of 1000 layers of pure linear transformations can be reproduced by a single layer. This is why non-linear activation functions are so important in deep learning.

5.3.10. Layer

We have been using the word layer quite often. Put simply, a layer is the highest-level building block in a DL architecture.

A layer is a container that usually receives weighted input, transforms it with a set of mostly non-linear functions and then passes these values as output to the next layer.

A layer is usually *uniform*, that is it contains one type of activation function, pooling, convolution etc. This can simplify it being compared to other parts of the network. The first and last layers in a network are called ***input*** and ***output*** layers, respectively, and all layers in between are called ***hidden*** layers.

5.4. CONVOLUTIONAL DEEP LEARNING

5.4.1. Convolution

People with a mathematical background usually understand what a convolution is all about. Let us take a layman's approach to understand it in the light of DL.

Basically, convolution is a mathematical operation which describes a rule of how to mix two functions or pieces of information. In the field of DL it is quite easy to understand. Let us break into parts to get a full picture. A convolution works by:

(1) taking input data and creating a *feature map*

(2) the convolution kernel mixes the input and

(3) outputs a transformed feature map.

Convolution is often interpreted as a ***filter***, where the *kernel* filters the feature map for information of a certain kind. Convolution is important in physics and mathematics and is often employed to bridge spatial and time domains and the frequency domain through the *convolution theorem.* This bridge is defined by the use of Fourier transforms. When you use a Fourier transform (FT) on both the kernel and the feature map, then the convolution operation is simplified significantly (integration becomes mere multiplication). Some of the fastest GPU implementations of convolutions (for example some implementations in the NVIDIA cuDNN library) currently make use of Fourier transforms. An example is shown below (Figure 5.2):

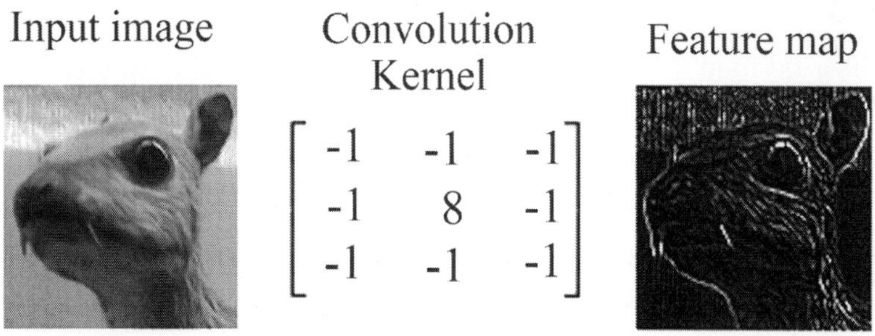

Figure 5.2: Convolution of an image with an edge detector convolution kernel.

It certainly is *unknown* which interpretation of convolution is correct for deep learning, the cross-correlation interpretation is currently the most useful. An example is shown below from image processing.

Convolution can be transformed to cross-correlation by reversing the kernel (upside-down image). The kernel can then be interpreted as a feature detector where a detected feature results in large outputs (white) and small outputs if no feature is present (black).

5.4.2. Pooling/Subsampling

Pooling is essentially a subsampling procedure, which takes input over a certain area and reduces it to a single value. In CNNs, this information concentration has a useful property which is that outgoing connections receive similar information. In other words, the information is "funneled" into the right place for the input feature map of the next convolutional layer. The larger the size of the pooling area, the more information is condensed, which leads to slim networks that fit more easily into GPU memory. However, if the pooling area is too large, too much information is thrown away and predictive performance decreases.

5.4.3. Convolutional Neural Network (CNN)

Again, in a layman's language a convolutional neural network (CNN), or preferably *convolutional network* or *convolutional net* uses convolutional layers that filter inputs for useful information. These convolutional layers have parameters that are learned so that these filters are adjusted automatically to extract the most useful information for the task at hand. Usually, multiple convolutional layers are used that filter information for more and more abstract information after each layer.

Convolutional networks usually also use *pooling layers* (see pooling). Pooling also reduces the memory consumption and thus allows for the usage of more convolutional layers. More recent convolutional networks use *inception modules* which employ 1×1 convolutional kernels to reduce the memory consumption further while speeding up the computation (and thus training). We illustrate these concepts pictorially in the Figure 5.3 given below.

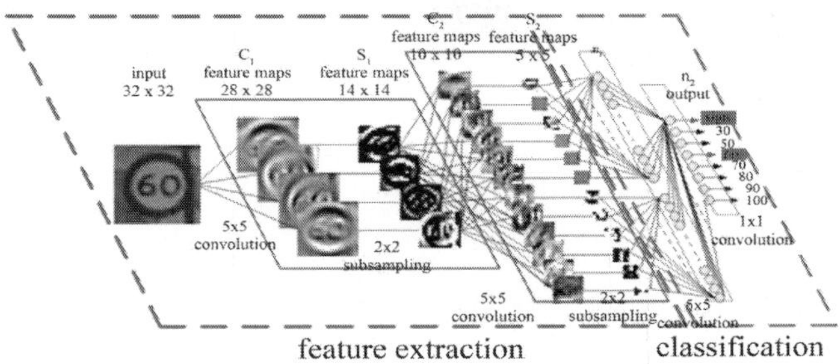

Figure 5.3: An image of a traffic sign is filtered by 4 5×5 convolutional kernels which create 4 feature maps, these feature maps are subsampled by max pooling.

5.5. AN EXAMPLE – MULTILAYER PERCEPTRONS

In this section we'll go over the basics of fitting neural networks in python, using *keras* and *tensorflow*. You'll need to have them both installed to be able to run this tutorial. I would recommend installation in the conda environment. Key take aways are to become familiar with the keras API and nomenclature and understand how MLPs relate to logistic regression.

```
In [2]:
%matplotlib inline

import matplotlib
import numpy as np
import matplotlib.pyplot as plt
plt.style.use('fivethirtyeight')

from sklearn.datasets import make_circles

from keras.layers import Input, Dense
from keras.models import Sequential

X, y = make_circles(n_samples=5000, factor=.3,
```

```
noise=.05)
X_train = X[:4000]
y_train = y[:4000]

X_val = X[4000:]
y_val = y[4000:]
num_variables = X.shape[1]
Using TensorFlow backend.
In [3]:
plt.scatter(X[:,0],      X[:,1],       c=y,cmap=plt.
cm.coolwarm)
plt.title("Classification Dataset")
plt.xlabel('x1')
plt.ylabel('x2')
plt.show()
```

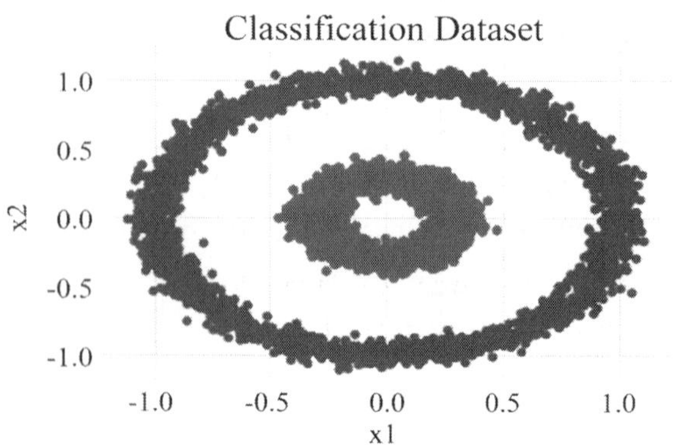

5.6. A FIRST MODEL WITH LOGISTIC REGRESSION

The first thing we have to do is to set up an input tensor. One the key advantages of using keras is that you only have to specify the shapes for the inputs and outputs, the shapes for the remaining tensors will be inferred automatically.

In our case, each input vector $x_i = <x_{i1}, x_{i2}>$ has two dimensions and the output is a single binary variable $y_i \in \{0,1\}$. Let's see how we would set this up with keras:

```
In [4]:
logreg = Sequential()
logreg.add(Dense(output_dim=1, input_dim=num_vari-
ables, activation='sigmoid'))
```

The first line tells keras to initialize a new **sequential** model. Sequential models take a single input tensor and produce a single output tensor. For the purposes of this tutorial, we're going to stick with the sequential model because it will have all of the functionality we'll need.

The bulk of the action happens on the second line, and it's a little terse so let's unpack it:

logreg.add tells keras that we want to add a new layer to our network.

Dense specifies that we want to use a fully-connected layer, aka a dense layer and is probably most opaque aspect of the keras API. In general, what a dense layer does create several different linear combinations of the output of the previous layer, followed by an element-wise application of an **activation function**. The number of linear combinations is specified by the number of 'hidden units' in the layer, which in this case is specified by output_dim.

input_dim=num_variables – Since this is the first layer in our network, we have to tell keras how big the input will be. In this case

activation='sigmoid' specifies which element-wise transformation we should apply the resulting output. Here we've specified a sigmoid transformation,

which is given by $\frac{1}{1+\exp(-t)}$, where t is the result of the linear combination. This function 'squashes' the weighted sum to be a number between 0 and 1.

Putting that all together, we are telling keras that we'd like a single linear combination of the input variables, followed by a sigmoid transformation, which will represent the probability that an observation belongs to the class y = 1. In this case, the **Dense** layer will have 2 parameters/weights $\{w_1, w_2\}$ and intercept/bias term since this is included by default.

Since this is a relatively simple case, we can write out the actual function implied by this model. It looks like this:

$$1+\exp^{-(w_0+w_1*x_{i1}+w_2*x_{i2})}$$

As you might be able to see now, this is equivalent to a logistic regression model, where we are modeling the log-odds as linear function of the input variables.

Enough talking, let's fit the model! That's easily enough accomplished in another 2 lines:

```
In [5]:
logreg.compile(optimizer='adam',loss='binary_cros-
sentropy', metrics=['accuracy'])
logreg.fit(X_train,     y_train,validation_data=[X_
val,y_val],verbose=0)
val_score = logreg.evaluate(X_val,y_val,verbose=0)
print "Accuracy on validation data: " + str(val_
score[1]*100) + "%"
Accuracy on validation data: 49.6%
In [6]:
# create a mesh to plot in
h = 0.02
x_min, x_max = X_val[:, 0].min(), X_val[:, 0].max()
y_min, y_max = X_val[:, 1].min(), X_val[:, 1].max()
xx, yy = np.meshgrid(np.arange(x_min, x_max, h),
                     np.arange(y_min, y_max, h))
Z = logreg.predict(np.c_[xx.ravel(), yy.ravel()])
Z = Z.reshape(xx.shape)
plt.contourf(xx, yy, Z, cmap=plt.cm.coolwarm, al-
pha=0.8)
plt.scatter(X_val[:,  0],  X_val[:,  1],  c=y_val,
cmap=plt.cm.coolwarm)
plt.xlim(xx.min(), xx.max())
plt.ylim(yy.min(), yy.max())
plt.title("Logistic Regression Decision Regions")
plt.show()
```

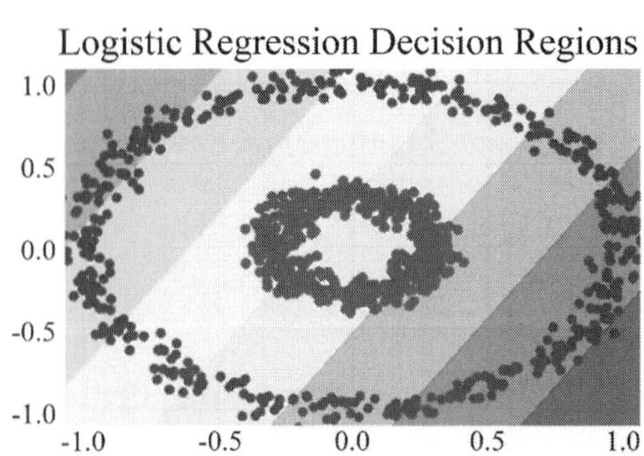

```
In [7]:
num_hidden = 5
mlp = Sequential()
mlp.add(Dense(output_dim=num_hidden,        input_
dim=num_variables, activation='relu'))
mlp.add(Dense(output_dim=1, activation='sigmoid'))
In [10]:
mlp.compile(optimizer='adam',
              loss='binary_crossentropy',
              metrics=['accuracy'])
mlp.fit(X_train,  y_train,validation_data=[X_val,y_
val],nb_epoch=20,verbose=0)
Out[10]:
<keras.callbacks.History at 0x7f9440155b50>
In [11]:
# create a mesh to plot in
h = 0.02
x_min, x_max = X_val[:, 0].min(), X_val[:, 0].max()
y_min, y_max = X_val[:, 1].min(), X_val[:, 1].max()
xx, yy = np.meshgrid(np.arange(x_min, x_max, h),
                     np.arange(y_min, y_max, h))
Z = mlp.predict(np.c_[xx.ravel(), yy.ravel()])
Z = Z.reshape(xx.shape)
plt.contourf(xx, yy, Z, cmap=plt.cm.coolwarm, al-
pha=0.8)
plt.scatter(X_val[:,  0],  X_val[:,  1],  c=y_val,
cmap=plt.cm.coolwarm)
plt.xlim(xx.min(), xx.max())
plt.ylim(yy.min(), yy.max())
plt.title("MLP with " + str(num_hidden) + " Hidden
Units")
plt.show()
```

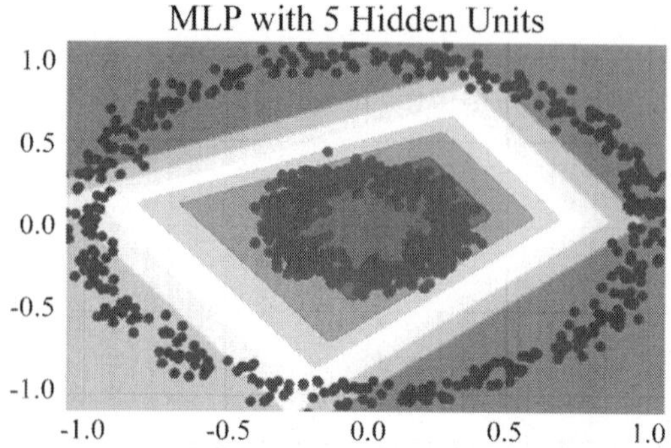

We can see that a neural network with 5 hidden units is trying to stitch together a relatively boxy set of lines to create piecewise linear functions in an effort to classify the points.

```
In [12]:
num_hidden = 128
mlp = Sequential()
mlp.add(Dense(output_dim=num_hidden,         input_
dim=num_variables, activation='relu'))
mlp.add(Dense(output_dim=1, activation='sigmoid'))
mlp.compile(optimizer='adam',
              loss='binary_crossentropy',
              metrics=['accuracy'])
mlp.fit(X_train,  y_train,validation_data=[X_val,y_
val],nb_epoch=10,verbose=1)
Train on 4000 samples, validate on 1000 samples
Epoch 1/10
4000/4000 [==============================] - 0s -
loss: 0.6049 - acc: 0.6467 - val_loss: 0.5108 -
val_acc: 0.9810
Epoch 2/10
4000/4000 [==============================] - 0s -
loss: 0.4036 - acc: 0.9995 - val_loss: 0.2923 -
val_acc: 1.0000
Epoch 3/10
4000/4000 [==============================] - 0s -
loss: 0.2100 - acc: 1.0000 - val_loss: 0.1412 -
val_acc: 1.0000
```

```
Epoch 4/10
4000/4000 [==============================] - 0s -
loss: 0.1048 - acc: 1.0000 - val_loss: 0.0740 -
val_acc: 1.0000
Epoch 5/10
4000/4000 [==============================] - 0s -
loss: 0.0584 - acc: 1.0000 - val_loss: 0.0439 -
val_acc: 1.0000
Epoch 6/10
4000/4000 [==============================] - 0s -
loss: 0.0364 - acc: 1.0000 - val_loss: 0.0286 -
val_acc: 1.0000
Epoch 7/10
4000/4000 [==============================] - 0s -
loss: 0.0247 - acc: 1.0000 - val_loss: 0.0200 -
val_acc: 1.0000
Epoch 8/10
4000/4000 [==============================] - 0s -
loss: 0.0177 - acc: 1.0000 - val_loss: 0.0147 -
val_acc: 1.0000
Epoch 9/10
4000/4000 [==============================] - 0s -
loss: 0.0133 - acc: 1.0000 - val_loss: 0.0112 - val_
acc: 1.0000 - ETA: 0s - loss: 0.0136 - acc: 1.0000
Epoch 10/10
4000/4000 [==============================] - 0s -
loss: 0.0103 - acc: 1.0000 - val_loss: 0.0088 -
val_acc: 1.0000
Out[12]:
<keras.callbacks.History at 0x7f9468c25f90>
In [13]:
# create a mesh to plot in
h = 0.02
x_min, x_max = X_val[:, 0].min(), X_val[:, 0].max()
y_min, y_max = X_val[:, 1].min(), X_val[:, 1].max()
xx, yy = np.meshgrid(np.arange(x_min, x_max, h),
                     np.arange(y_min, y_max, h))
Z = mlp.predict(np.c_[xx.ravel(), yy.ravel()])
Z = Z.reshape(xx.shape)
```

```
plt.contourf(xx, yy, Z, cmap=plt.cm.coolwarm, al-
pha=0.8)
plt.scatter(X_val[:, 0], X_val[:, 1], c=y_val,
cmap=plt.cm.coolwarm)
plt.xlim(xx.min(), xx.max())
plt.ylim(yy.min(), yy.max())
plt.title("MLP with " + str(num_hidden) + " Hidden
Units")
plt.show()
```

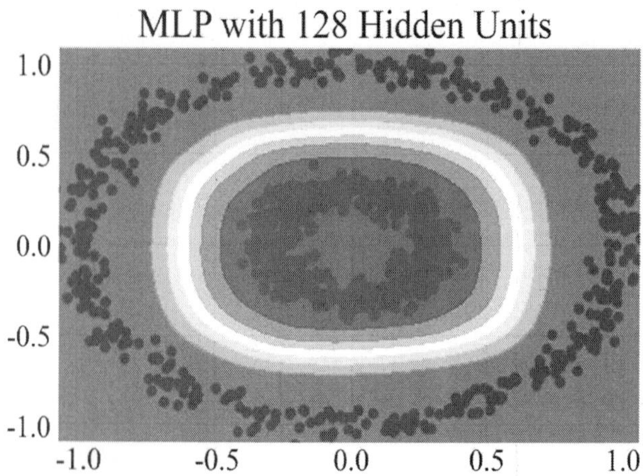

Not that we need to for this example, but we can easily extend this model to add more layers, simply by adding another dense layer.

```
In [14]:
mlp = Sequential()
mlp.add(Dense(output_dim=num_hidden, input_
dim=num_variables, activation='relu'))
mlp.add(Dense(output_dim=num_hidden,
activation='relu'))
mlp.add(Dense(output_dim=1, activation='sigmoid'))
mlp.compile(optimizer='adam',
              loss='binary_crossentropy',
              metrics=['accuracy'])
mlp.fit(X_train, y_train,validation_data=[X_val,y_
val],nb_epoch=10,verbose=1)
Train on 4000 samples, validate on 1000 samples
```

```
Epoch 1/10
4000/4000 [==============================] - 0s -
loss: 0.3127 - acc: 0.8510 - val_loss: 0.0372 -
val_acc: 1.0000
Epoch 2/10
4000/4000 [==============================] - 0s -
loss: 0.0112 - acc: 1.0000 - val_loss: 0.0032 -
val_acc: 1.0000
Epoch 3/10
4000/4000 [==============================] - 0s -
loss: 0.0020 - acc: 1.0000 - val_loss: 0.0012 - val_
acc: 1.0000 - ETA: 0s - loss: 0.0025 - acc: 1.0000
- ETA: 0s - loss: 0.0022 - acc: 1.0000
Epoch 4/10
4000/4000 [==============================] - 0s -
loss: 8.6576e-04 - acc: 1.0000 - val_loss: 5.9462e-
04 - val_acc: 1.0000
Epoch 5/10
4000/4000 [==============================] - 0s -
loss: 4.9211e-04 - acc: 1.0000 - val_loss: 3.6253e-
04 - val_acc: 1.0000 - ETA: 0s - loss: 5.1859e-04
- acc: 1.0000
Epoch 6/10
4000/4000 [==============================] - 0s -
loss: 3.1913e-04 - acc: 1.0000 - val_loss: 2.4520e-
04 - val_acc: 1.0000 - ETA: 0s - loss: 3.3042e-04
- acc: 1.0000
Epoch 7/10
4000/4000 [==============================] - 0s -
loss: 2.2346e-04 - acc: 1.0000 - val_loss: 1.7577e-
04 - val_acc: 1.0000
Epoch 8/10
4000/4000 [==============================] - 0s -
loss: 1.6531e-04 - acc: 1.0000 - val_loss: 1.3257e-
04 - val_acc: 1.0000 - ETA: 0s - loss: 1.8606e-04
- acc: 1.0000
Epoch 9/10
4000/4000 [==============================] - 0s -
loss: 1.2738e-04 - acc: 1.0000 - val_loss: 1.0297e-
```

```
04 - val_acc: 1.0000
Epoch 10/10
4000/4000 [==============================] - 0s -
loss: 1.0074e-04 - acc: 1.0000 - val_loss: 8.2025e-
05 - val_acc: 1.0000 - ETA: 0s - loss: 1.0449e-04
- acc: 1.0000
Out[14]:
<keras.callbacks.History at 0x7f9438402790>
In [16]:
h = 0.02
x_min, x_max = X_val[:, 0].min(), X_val[:, 0].max()
y_min, y_max = X_val[:, 1].min(), X_val[:, 1].max()
xx, yy = np.meshgrid(np.arange(x_min, x_max, h),
                     np.arange(y_min, y_max, h))
Z = mlp.predict(np.c_[xx.ravel(), yy.ravel()])
Z = Z.reshape(xx.shape)
plt.contourf(xx, yy, Z, cmap=plt.cm.coolwarm, al-
pha=0.8)
plt.scatter(X_val[:, 0], X_val[:, 1], c=y_val,
cmap=plt.cm.coolwarm)
plt.xlim(xx.min(), xx.max())
plt.ylim(yy.min(), yy.max())
plt.title("MLP with 2 Hidden Layers and " + str(num_
hidden) + " Units")
plt.show()
```

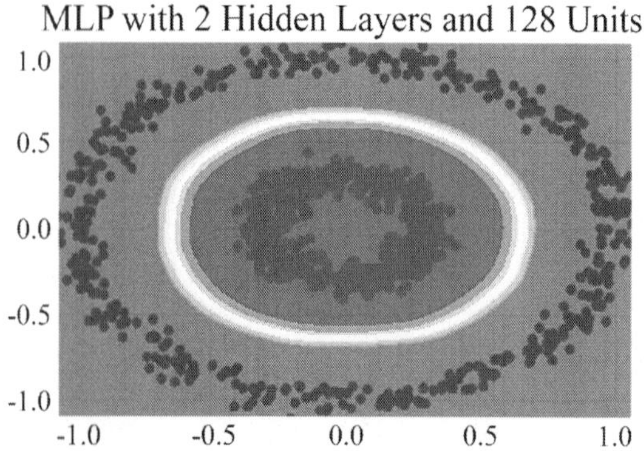

That's it for this tutorial. Hopefully it helped you get a handle on how to use the keras API to build some simple neural networks.

We will revisit all the basics covered in previous sections, but this time we will have a hands-on detour to the fundamentals. So gear up.

5.7. IMPLEMENTING BASICS IN DEEP LEARNING

Although it may sound repetitive, still we would like to revisit the basics in the previous sections. But, this time in a condensed form with computational exercises.

In machine learning and cognitive science, an artificial neural network (ANN) is a network inspired by biological neural networks which are used to estimate or approximate functions that can depend on a large number of inputs that are generally unknown. An ANN is built from nodes (neurons) stacked in layers between the feature vector and the target vector.

A node in a neural network is built from Weights and Activation function

An early version of ANN built from one node was called the **Perceptron.**

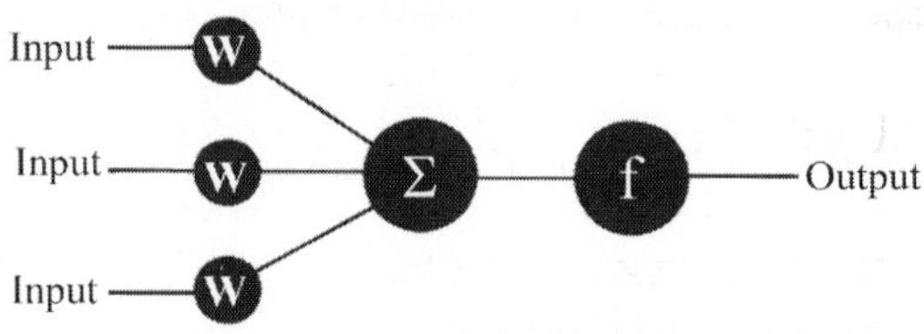

Perceptron

The Perceptron is an algorithm for supervised learning of binary classifiers. functions that can decide whether an input (represented by a vector of numbers) belongs to one class or another. Much like logistic regression, the weights in a neural net are being multiplied by the input vector summed up and feeded into the activation function's input. A Perceptron Network can be designed to have *multiple layers*, leading to the **Multi-Layer Perceptron** (aka MLP)

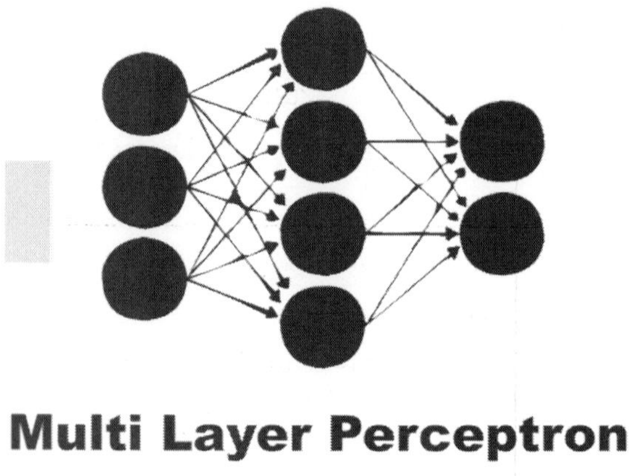

5.7.1. Single Layer Neural Network

We illustrate a single layer NN below:

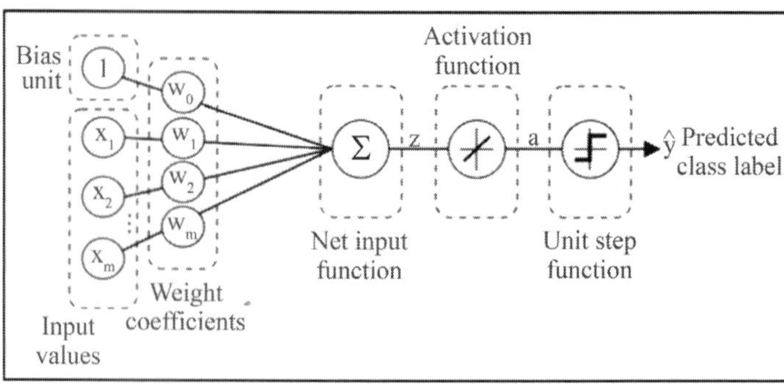

5.7.2. Weights Update Rule

- We use a **gradient descent** optimization algorithm to learn the *Weights Coefficients* of the model.
- In every **epoch** (pass over the training set), we update the weight vector using the following update rule:

$$w = w + \Delta w, where \Delta w = -\eta \Delta J(w)$$

In other words, we computed the gradient based on the whole training set and updated the weights of the model by taking a step into the **opposite direction** of the gradient $\Delta J(w)$.

In order to fin the **optimal weights of the model**, we optimized an objective function (e.g., the Sum of Squared Errors (SSE)) cost function

$J(w)$.

Furthermore, we multiply the gradient by a factor, the learning rate η, which we choose carefully to balance the **speed of learning** against the risk of overshooting the global minimum of the cost function.

5.7.3. Gradient Descent

In **gradient descent optimization**, we update all the **weights simultaneously** after each epoch, and we define the *partial derivative* for each weight in the weight vector as follows:

$$\frac{\partial}{\partial w_j} J(w) = \sum_i (y^{(i)} - a^{(i)}) x_j^{(i)}$$

Note: The superscript (i) refers to the i-th sample. The subscript j refers to the j-th dimension/feature

Here $y^{(i)}$ is the target class label of a particular sample $x^{(i)}$, and $a^{(i)}$ is the **activation** of the neuron

(which is a linear function in the special case of Perceptron).

We define the **activation function** $\phi(\cdot)$ as follows:

$$\phi(z) = z = a = \sum_j w_j x_j = w^T x$$

5.7.4. Binary Classification

While we used the **activation** $\phi(z)$ to compute the gradient update, we may use a **threshold function** *(Heaviside function)* to squash the continuous-valued output into binary class labels for prediction:

$$\hat{y}\begin{cases} 1 & if\, \phi(z) \geq 0 \\ 0 & otherwise \end{cases}$$

5.7.5. Building Neural Nets from scratch

We will build the neural networks from first principles. We will create a very simple model and understand how it works. We will also be implementing backpropagation algorithm. This is for instructive purpose – for us to understand how ANN works. Libraries like theano have highly optimized code.

5.7.6. Perceptron and Adaline Models

If you want a sneak peek of alternate (production ready) implementation of *Perceptron* for instance try:

from sklearn.linear_model import Perceptron

5.7.7. Introducing the Multi-Layer Neural Network Architecture

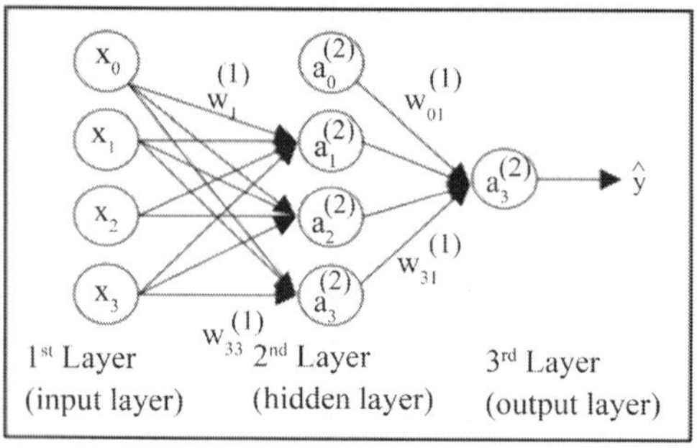

Now we will see how to connect **multiple single neurons** to a **multi-layer feedforward neural network**; this special type of network is also called a **multi-layer perceptron** (MLP). The figure shows the concept of an **MLP** consisting of three layers: one *input* layer, one *hidden* layer, and one *output* layer. The units in the hidden layer are fully connected to the input layer, and the output layer is fully connected to the hidden layer, respectively. If such a network has **more than one hidden layer**, we also call it a **deep artificial neural network**.

5.7.8. Notation

We denote the ith activation unit in the lth layer as $a_i^{(1)}$, and the activation units $a_0^{(1)}$ and $a_0^{(1)}$ are the **bias units**, respectively, which we set equal to . The *activation* of the units in the **input layer** is just its input plus the bias unit:

$$a^{(1)} = [a_0^{(1)}, a_0^{(1)}, \dots, a_m^{(1)}]^T = [1, x_1^{(i)}, \dots, x_m^{(i)}]^T$$

Note: $x_j^{(j)}$ refers to the jth feature/dimension of the ith sample

5.7.9. Notes on Notation (Usually) Adopted

The terminology around the indices (subscripts and superscripts) may look a little bit confusing at first. You may wonder why we wrote $w_{j,k}^{(l)}$ and not $w_{k,l}^{(l)}$ to refer to the **weight coefficient** that connects the *kth* unit in layer to the jth unit in layer $l+1$. What may seem a little bit quirky at first will make much more sense later when we **vectorize** the neural network representation. For example, we will summarize the weights that connect the input and hidden layer by a matrix

$W^{(1)} \in \mathbb{R}^{h \times [m+1]}$

where h is the number of hidden units and m+1 is the number of hidden units plus bias unit.

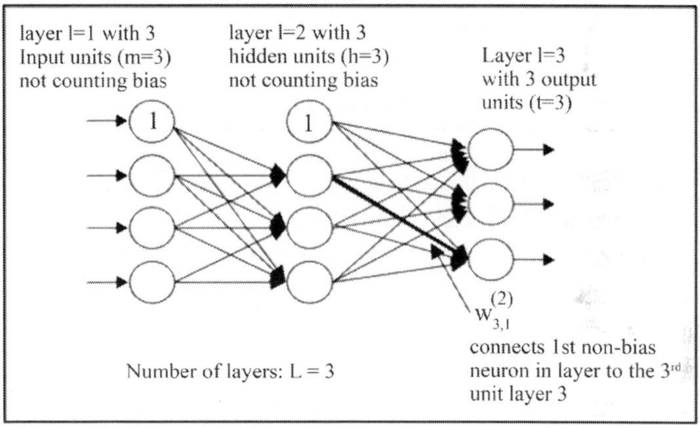

5.7.10. Forward Propagation

- Starting at the input layer, we forward propagate the patterns of the training data through the network to generate an output.
- Based on the network's output, we calculate the error that we want to minimize using a cost function that we will describe later.
- We backpropagate the error, find its derivative with respect to each weight in the network, and update the model.

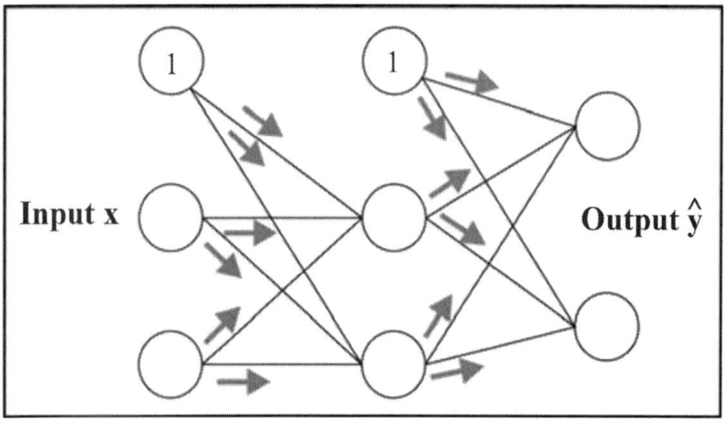

5.7.11. Backward Propagation

The weights of each neuron are learned by **gradient descent**, where each neuron's error is derived with respect to it's weight.

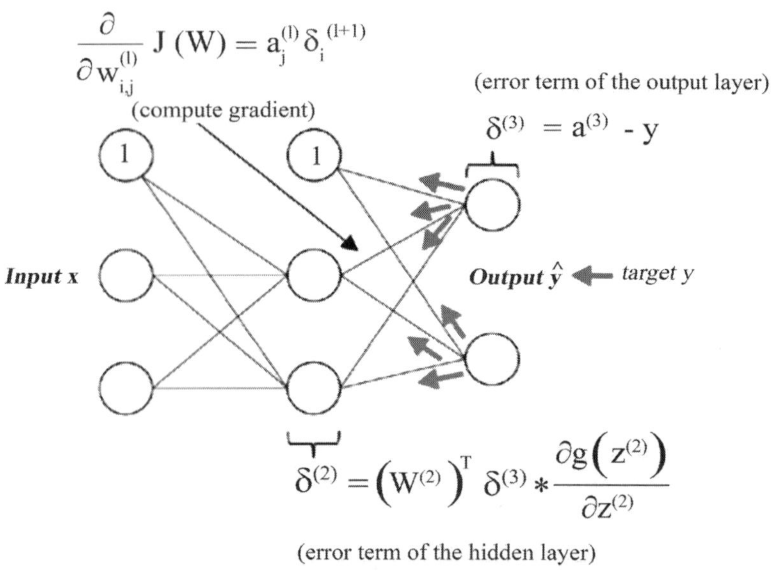

Optimization is done for each layer with respect to the previous layer in a technique known as **BackPropagation**.

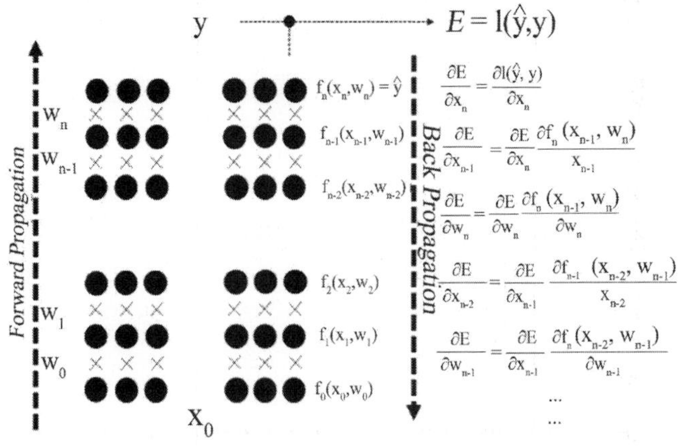

Let us now jump start with our model building from scratch. Open a Jupyter Notebook and get started. To access the sample dataset, the reader is referred to download the file at the following link below:

https://www.dropbox.com/s/haf6bptu4rmlrb8/intro_to_ann.csv?dl=0

```
In [1]:
# Import the required packages
import numpy as np
import pandas as pd
import matplotlib
import matplotlib.pyplot as plt
import scipy

In [2]:
# Display plots in notebook
%matplotlib inline
# Define plot's default figure size
matplotlib.rcParams['figure.figsize'] = (10.0, 8.0)
In [3]:
#read the datasets
train = pd.read_csv(."./data/intro_to_ann.csv")
In [4]:
X, y = np.array(train.ix[:,0:2]), np.array(train.
ix[:,2])
In [5]:
X.shape
```

```
Out[5]:
(500, 2)
In [6]:
y.shape
Out[6]:
(500,)
In [7]:
#Let's plot the dataset and see how it is
plt.scatter(X[:,0], X[:,1], s=40, c=y, cmap=plt.
cm.BuGn)
Out[7]:
<matplotlib.collections.PathCollection at
0x10e9329b0>
```

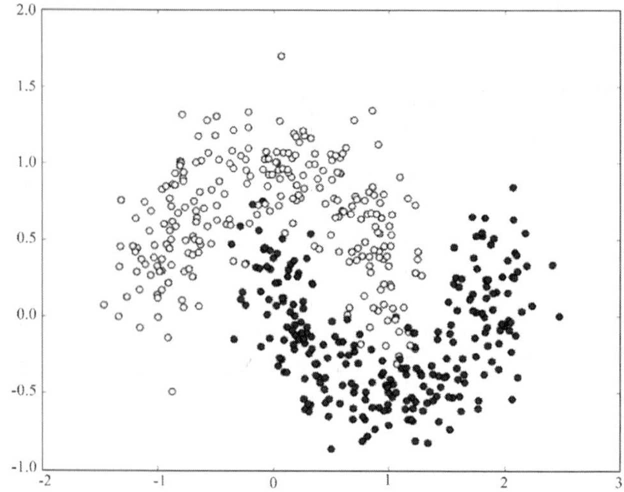

5.7.12. Start Building Our MLP Building Blocks

Note: This process will eventually result in our own Neural Networks class

```
In [8]:
import random
random.seed(123)

# calculate a random number where:  a <= rand < b
def rand(a, b):
    return (b-a)*random.random() + a
```

Function to generate a random number, given two numbers

5.7.13. Where Will it be Used?

When we initialize the neural networks, the weights have to be randomly assigned.

```
In [9]:
# Make a matrix
def makeMatrix(I, J, fill=0.0):
    return np.zeros([I,J])
```

Define our activation function. Let's use sigmoid function

```
In [10]:
# our sigmoid function
def sigmoid(x):
    #return math.tanh(x)
    return 1/(1+np.exp(-x))
```

Derivative of our activation function

We need this when we run the backpropagation algorithm

```
In [11]:
# derivative of our sigmoid function, in terms of
the output (i.e. y)
def dsigmoid(y):
    return y - y**2
```

When we first create a neural networks architecture, we need to know the number of inputs, number of hidden layers and number of outputs. The weights have to be randomly initialized.

```
class MLP:
    def __init__(self, ni, nh, no):
        # number of input, hidden, and output nodes
        self.ni = ni + 1 # +1 for bias node
        self.nh = nh
        self.no = no

        # activations for nodes
        self.ai = [1.0]*self.ni
        self.ah = [1.0]*self.nh
        self.ao = [1.0]*self.no

        # create weights
        self.wi = makeMatrix(self.ni, self.nh)
```

```
        self.wo = makeMatrix(self.nh, self.no)

        # set them to random vaules
      self.wi = rand(-0.2, 0.2, size=self.wi.shape)
      self.wo = rand(-2.0, 2.0, size=self.wo.shape)

        # last change in weights for momentum
        self.ci = makeMatrix(self.ni, self.nh)
        self.co = makeMatrix(self.nh, self.no)
```

Let us define our activation function in a classs.

```
def activate(self, inputs):

    if len(inputs) != self.ni-1:
        print(inputs)
        raise ValueError('wrong number of inputs')

    # input activations
    for i in range(self.ni-1):
        self.ai[i] = inputs[i]

    # hidden activations
    for j in range(self.nh):
        sum_h = 0.0
        for i in range(self.ni):
            sum_h += self.ai[i] * self.wi[i][j]
        self.ah[j] = sigmoid(sum_h)

    # output activations
    for k in range(self.no):
        sum_o = 0.0
        for j in range(self.nh):
            sum_o += self.ah[j] * self.wo[j][k]
        self.ao[k] = sigmoid(sum_o)

    return self.ao[:]
```

BackPropagation

```
def backPropagate(self, targets, N, M):

    if len(targets) != self.no:
```

```
        print(targets)
          raise ValueError('wrong number of target
values')

    # calculate error terms for output
    output_deltas = np.zeros(self.no)
    for k in range(self.no):
        error = targets[k]-self.ao[k]
         output_deltas[k] = dsigmoid(self.ao[k]) *
error

    # calculate error terms for hidden
    hidden_deltas = np.zeros(self.nh)
    for j in range(self.nh):
        error = 0.0
        for k in range(self.no):
           error += output_deltas[k]*self.wo[j][k]
         hidden_deltas[j] = dsigmoid(self.ah[j]) *
error

    # update output weights
    for j in range(self.nh):
        for k in range(self.no):
            change = output_deltas[k] * self.ah[j]
            self.wo[j][k] += N*change +
                             M*self.co[j][k]
            self.co[j][k] = change

    # update input weights
    for i in range(self.ni):
        for j in range(self.nh):
            change = hidden_deltas[j]*self.ai[i]
            self.wi[i][j] += N*change +
                             M*self.ci[i][j]
            self.ci[i][j] = change

    # calculate error
    error = 0.0
    for k in range(len(targets)):
```

```
        error += 0.5*(targets[k]-self.ao[k])**2
    return error
In [12]:
# Putting all together

class MLP:
    def __init__(self, ni, nh, no):
        # number of input, hidden, and output nodes
        self.ni = ni + 1 # +1 for bias node
        self.nh = nh
        self.no = no

        # activations for nodes
        self.ai = [1.0]*self.ni
        self.ah = [1.0]*self.nh
        self.ao = [1.0]*self.no

        # create weights
        self.wi = makeMatrix(self.ni, self.nh)
        self.wo = makeMatrix(self.nh, self.no)

        # set them to random vaules
        for i in range(self.ni):
            for j in range(self.nh):
                self.wi[i][j] = rand(-0.2, 0.2)
        for j in range(self.nh):
            for k in range(self.no):
                self.wo[j][k] = rand(-2.0, 2.0)

        # last change in weights for momentum
        self.ci = makeMatrix(self.ni, self.nh)
        self.co = makeMatrix(self.nh, self.no)


    def backPropagate(self, targets, N, M):

        if len(targets) != self.no:
            print(targets)
            raise ValueError('wrong number of tar-
```

```
get values')

        # calculate error terms for output
        output_deltas = np.zeros(self.no)
        for k in range(self.no):
            error = targets[k]-self.ao[k]
           output_deltas[k] = dsigmoid(self.ao[k])
* error

        # calculate error terms for hidden
        hidden_deltas = np.zeros(self.nh)
        for j in range(self.nh):
            error = 0.0
            for k in range(self.no):
               error += output_deltas[k]*self.wo[j]
[k]
           hidden_deltas[j] = dsigmoid(self.ah[j])
* error

        # update output weights
        for j in range(self.nh):
            for k in range(self.no):
                 change = output_deltas[k] * self.
ah[j]
               self.wo[j][k] += N*change + M*self.
co[j][k]
                self.co[j][k] = change

        # update input weights
        for i in range(self.ni):
            for j in range(self.nh):
               change = hidden_deltas[j]*self.ai[i]
                self.wi[i][j] += N*change + M*self.
ci[i][j]
                self.ci[i][j] = change

        # calculate error
        error = 0.0
        for k in range(len(targets)):
```

```
            error += 0.5*(targets[k]-self.ao[k])**2
        return error
    def test(self, patterns):
           self.predict  =  np.empty([len(patterns),
self.no])
        for i, p in enumerate(patterns):
            self.predict[i] = self.activate(p)
            #self.predict[i] = self.activate(p[0])

    def activate(self, inputs):

        if len(inputs) != self.ni-1:
            print(inputs)
             raise ValueError('wrong number of in-
puts')

        # input activations
        for i in range(self.ni-1):
            self.ai[i] = inputs[i]

        # hidden activations
        for j in range(self.nh):
            sum_h = 0.0
            for i in range(self.ni):
                sum_h += self.ai[i] * self.wi[i][j]
            self.ah[j] = sigmoid(sum_h)

        # output activations
        for k in range(self.no):
            sum_o = 0.0
            for j in range(self.nh):
                sum_o += self.ah[j] * self.wo[j][k]
            self.ao[k] = sigmoid(sum_o)

        return self.ao[:]


      def  train(self,  patterns,  iterations=1000,
N=0.5, M=0.1):
```

```
        # N: learning rate
        # M: momentum factor
        patterns = list(patterns)
        for i in range(iterations):
            error = 0.0
            for p in patterns:
                inputs = p[0]
                targets = p[1]
                self.activate(inputs)
             error += self.backPropagate([targets],
N, M)
            if i % 5 == 0:
                  print('error in interation %d :
%-.5f' % (i,error))
             print('Final training error: %-.5f' %
error)
```

5.7.14. Running the Model on our Dataset

```
In [13]:
# create a network with two inputs, one hidden, and
one output nodes
ann = MLP(2, 1, 1)

%timeit -n 1 -r 1 ann.train(zip(X,y), iterations=2)
error in interation 0 : 53.62995
Final training error: 53.62995
Final training error: 47.35136
1 loop, best of 1: 36.7 ms per loop
```

5.7.15. Predicting on Training Dataset and Measuring in-Sample Accuracy

```
In [14]:
%timeit -n 1 -r 1 ann.test(X)
1 loop, best of 1: 11.8 ms per loop
In [15]:
prediction     =     pd.DataFrame(data=np.array([y,
np.ravel(ann.predict)]).T,
                           columns=["actual", "pre-
diction"])
```

```
prediction.head()
Out[15]:
```

	ac - tual	prediction
0	1.0	0.491100
1	1.0	0.495469
2	0.0	0.097362
3	0.0	0.400006
4	1.0	0.489664

In [16]:

```
np.min(prediction.prediction)
Out[16]:
0.07655307811318012 9
```

Let's visualize and observe the results

```
In [17]:
# Helper function to plot a decision boundary.
# This generates the contour plot to show the deci-
sion boundary visually
def plot_decision_boundary(nn_model):
    # Set min and max values and give it some padding
    x_min, x_max = X[:, 0].min() - .5, X[:, 0].max()
+ .5
    y_min, y_max = X[:, 1].min() - .5, X[:, 1].max()
+ .5
     h = 0.01
     # Generate a grid of points with distance h be-
tween them
    xx, yy = np.meshgrid(np.arange(x_min, x_max, h),
                         np.arange(y_min, y_max, h))
     # Predict the function value for the whole gid
     nn_model.test(np.c_[xx.ravel(), yy.ravel()])
     Z = nn_model.predict
     Z[Z>=0.5] = 1
     Z[Z<0.5] = 0
     Z = Z.reshape(xx.shape)
     # Plot the contour and training examples
     plt.contourf(xx, yy, Z, cmap=plt.cm.Spectral)
```

```
    plt.scatter(X[:, 0], X[:, 1], s=40,  c=y,
cmap=plt.cm.BuGn)
In [18]:
plot_decision_boundary(ann)
plt.title("Our initial model")
Out[18]:
<matplotlib.text.Text at 0x110bafd68>
```

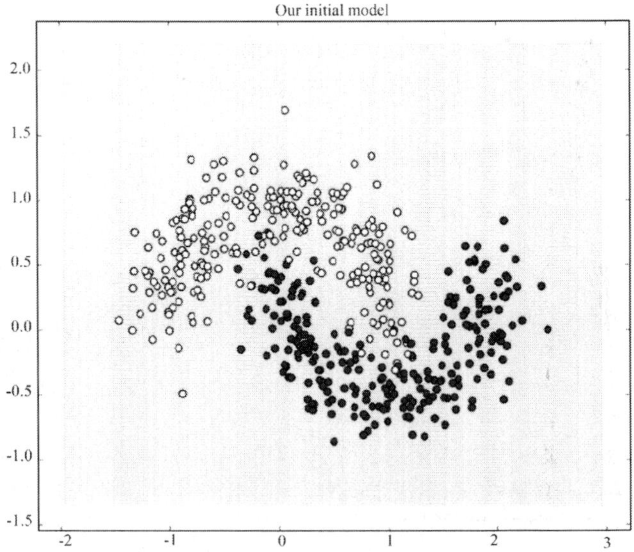

6 CHAPTER

PATIENT CARE: A PRACTICAL COMPUTATIONAL APPROACH

CONTENTS

Everyday, the healthcare industry creates large amounts of patient and clinical data and stores them in electronic health records. Most of this data has previously been inaccessible, in part due to patient privacy concerns, which poses a challenge to researchers working on the analysis of health records.

However, initiatives like the Medical Information Mart For Intensive Care (MIMIC) database project have allowed for everyone to use and experiment with health data. In particular, the MIMIC database is a critical care database made freely available for researchers around the world to develop and evaluate intensive care unit (ICU) patient monitoring and decision support systems that will improve the efficiency, accuracy and timeliness of clinical decision-making in critical care. The database can be accessed at:

https://mimic.physionet.org/

Using the MIMIC database which includes vital signs, medications, diagnostic code and many more variables, along with feature engineering techniques, one can investigate whether one can build a robust classifier to perform the following tasks which can be used to help hospitals and care units make decisions about the management of resources and staffing:

1. ***Mortality prediction***: Patient mortality is a primary outcome of interest in hospital settings, therefore early detection of at-risk patients is important. Mortality prediction is formulated as a binary classification problem, where the target label is whether the patient died before hospital discharge, or within a fixed time window e.g. likelihood of dying in 1 day, 1 week or 1 month in time.
2. ***Hospital length of stay***: Forecasting the hospital length of stay of a patient is also an important problem that could help manage hospital resources and cost. This can be formulated as a regression task where the amount of days that a patient is likely to spend at the hospital is predicted.

Few logical and practical approaches are possible. We present one such approach in this chapter. One can first experiment with a basic model which ignores temporal structure in the data, and attempts to perform mortality prediction. Briefly, for the first model, one can attempt to complete the following tasks:

1. Extract predictor variables from the MIMIC database
2. Build machine learning models for the predictions tasks

3. Evaluate and compare the performance of various algorithms

In the second improved model, one takes into account the complexity of time series data, since past vital signs and how they previously fluctuated can be strong indicators of the current condition of a patient. The improved model of patient mortality is still a binary classification, but one can now incorporate temporal features from the time series of the vital signs data. For example, one can add information about the health state of the patient from previous points in time e.g. the previous hours, the mean/median/min/max value of the last 6 hours, etc.

One can explore the pros and cons of each model, the challenges of using time series data and compare the different approaches.

6.1. OVERVIEW OF MIMIC

The MIMIC database mainly includes demographic, administrative, clinical data and much more from thousands of critical care patients. The table and the plot below provides basic descriptive statistics of the patients and an overview of the dataset.

Information	***Totals***
Age, years, median data-cleaning	65.769
Gender, male (%)	56.207
Distinct number of patients	38,597
Distinct ICU stays	53,423
Hospital admissions	49,785
Hospital length of stay (days)	11.545
ICU length of stay (days)	2.144
Hospital mortality (%)	11.545
ICU mortality (%)	8.545

6.2. MIMIC DATABASE

The MIMIC database, short for Medical Information Mart For Intensive Care, is a database that holds a large amount of information relating to patients admitted to the intensive care unit (ICU) at a large care hospital. The dataset contains clinical data (demographics, diagnoses, laboratory values), time-stamped nurse-verified physiological measurements, documented progress notes by care providers, and much more. This rich information that is available allows to research, develop and evaluate intensive care unit (ICU) patient monitoring and decision support systems that have the potential to improve the efficiency and timeliness of clinical decision-making in critical care.

Because of the large size of data, it is loaded in a database system like PostgreSQL to better manage and handle the high quantity of information. Before we dive into any machine learning or data analysis, we'll first explore the contents of the dataset using SQL and different Python libraries for visualization and analysis. More specifically, we'll cover the following in this notebook:

- How to select data from a database using Structured Query Language (SQL)
- Query basic demographics from the MIMIC database
- Plot histograms displaying patient information
- Create descriptive statistics describing the demographics of the dataset
- Extract vital signs and laboratory measurements for a single patient

6.2.1. Pre-requisites

Although access to the database is freely available to researchers around the world, a Human Subjects training is required to handle sensitive patient information. More details for the formal request can be found in the following:

https://mimic.physionet.org/gettingstarted/access/

The following code will assume a prior installation of the MIMIC database in a local Postgres database. The installation procedure can be followed at the following link:

https://mimic.physionet.org/tutorials/install-mimic-locally-ubuntu/

A tutorial introducing basic SQL concepts with the MIMIC-III database can also be found in the Appendix to this book.

6.2.2. Exploring the MIMIC Database

Since MIMIC is a relational database containing tables of patient data who stayed at the ICU, data will be extracted using SQL queries and then stored as *Pandas* dataframes for easy manipulation. After setting up the Postgres database, we'll use the *psycopg2* library to connect to the SQL database for queries.

```
In [1]:
import numpy as np
import pandas as pd
import matplotlib.pyplot as plt
import psycopg2
import getpass
%matplotlib inline
plt.style.use('ggplot')
In [5]:
# SQL database config
sqluser = ''
dbname = 'MIMIC3'
schema_name = 'mimiciii'
hostname = ''
port = 5432
pwd = getpass.getpass()
........
In [6]:
# Connect to local postgres version of mimic
con = psycopg2.connect(dbname=dbname, user=sqluser,
host=hostname, port=5432, password=pwd)
cur = con.cursor()
cur.execute('SET search_path to ' + schema_name)
```

A quick overview of MIMIC database is given below. The tables are linked by identifiers with the suffix ID: SUBJECT_ID which represents a unique patient, HADM_ID which represents a unique patient hospital stay and ICUSTAY_ID which represents a unique patient ICU stay.

The following tables are used to define and track patient stays:

- ADMISSIONS: Every unique hospitalization for each patient in the database (defines HADM_ID)
- CALLOUT: Information regarding when a patient was cleared for ICU discharge and when the patient was actually discharged
- ICUSTAYS: Every unique ICU stay in the database (defines ICUSTAY_ID)
- PATIENTS: Every unique patient in the database (defines

SUBJECT_ID)

- SERVICES: The clinical service under which a patient is registered
- TRANSFERS: Patient movement from bed to bed within the hospital, including ICU admission and discharge

A simple SQL query can therefore be used as follows, which would select only female subjects from the database, or patients with the subject_id 109 or 127.

```
SELECT *
FROM patients
WHERE subject_id = 109
OR subject_id = 127;
OR gender = 'F'
```

6.2.3. Joining Admissions, Icustays, and Patients Tables

We'll now combine multiple tables in order obtain a more detailed description of the data. In particular, we'll extract the following columns that contain basic information regarding the patients in the database:

- Identifiers (subject_id, hadm_id, icustay_id)
- The ICU care unit
- The patients date of birth, death
- The time the patient entered and left the ICU
- The length of stay at the hospital
- The age during the ICU and the hospital stay
- Whether the patient died in the ICU or in the hospital

In [7]:

```
query = \
""""
WITH population as (
SELECT a.subject_id, a.hadm_id, i.icustay_id,
a.admittime as hosp_admittime, a.dischtime as hosp_dischtime,
i.first_careunit,
```

```
  DENSE_RANK() over(PARTITION BY a.hadm_id ORDER BY i.intime
ASC) as icu_seq,
  p.dob, p.dod, i.intime as icu_intime, i.outtime as icu_outtime,
  i.los as icu_los,
  round((EXTRACT(EPOCH FROM (a.dischtime-a.admittime))/60/60/24)
:: NUMERIC, 4) as hosp_los,
  p.gender,
round((EXTRACT(EPOCHFROM(a.admittime-p.dob))/60/60/24/365.242)
:: NUMERIC, 4) as age_hosp_in,
  round((EXTRACT(EPOCH FROM (i.intime-p.dob))/60/60/24/365.242)
:: NUMERIC, 4) as age_icu_in,
  hospital_expire_flag,
  CASE WHEN p.dod IS NOT NULL
    AND p.dod >= i.intime – interval '6 hour'
    AND p.dod <= i.outtime + interval '6 hour' THEN 1
    ELSE 0 END AS icu_expire_flag
   FROM admissions a
   INNER JOIN icustays i
   ON a.hadm_id = i.hadm_id
   INNER JOIN patients p
   ON a.subject_id = p.subject_id
   ORDER BY a.subject_id, i.intime
)
SELECT *
FROM population
WHERE age_hosp_in >= 16;
""""
query_output = pd.read_sql_query(query,con)
query_output.head()
Out[7]:
```

	sub-ject_id	hadm_id	icustay_id	hosp_ad-mittime	hosp_dis-chtime	first_ca-reunit	icu_seq	dob	dod	icu_in-time	icu_out-time	icu_los	hosp_los	gen-der	age_hosp_in	age_icu_in	hospi-tal_ex-pire_flag	icu_ex-pire_flag
0	3	145834	211552	2101-10-20 19:08:00	2101-10-31 13:58:00	MICU	1	2025-04-11	2102-06-14	2101-10-20 19:10:11	2101-10-26 20:43:09	6.0646	10.7847	M	76.5268	76.5268	0	**0**
1	4	185777	294638	2191-03-16 00:28:00	2191-03-23 18:41:00	MICU	1	2143-05-12	NaT	2191-03-16 00:29:31	2191-03-17 16:46:31	1.6785	7.7590	F	47.8450	47.8450	0	**0**
2	6	107064	228232	2175-05-30 07:15:00	2175-06-15 16:00:00	SICU	1	2109-06-21	NaT	2175-05-30 21:30:54	2175-06-03 13:39:54	3.6729	16.3646	F	65.9407	65.9423	0	**0**
3	9	150750	220597	2149-11-09 13:06:00	2149-11-14 10:15:00	MICU	1	2108-01-26	2149-11-14	2149-11-09 13:07:02	2149-11-14 20:52:14	5.3231	4.8813	M	41.7902	41.7902	1	**1**
4	11	194540	229441	2178-04-16 06:18:00	2178-05-11 19:00:00	SICU	1	2128-02-22	2178-11-14	2178-04-16 06:19:32	2178-04-17 20:21:05	1.5844	25.5292	F	50.1483	50.1483	0	0

Using the above columns, we can compute the following points for general statistics of the dataset:

- Hospital admissions, no. (% of total admissions)
- Distinct ICU stays, no. (% of total unit stays)
- Age, yrs, mean ± SD
- Gender, male, percent of unit stays
- ICU length of stay, median days (IQR)
- Hospital length of stay, median days (IQR)
- ICU mortality, percent of unit stays
- Hospital mortality, percent of unit stays

Note: the dates may seem futuristic, for example, born in 2191, but they are presented that way to protect patient privacy. Age and length of stay can still be computed from these dates.

```
In [8]:
len(query_output)
Out[8]:
53423
In [12]:
query_output.loc[query_output.age_icu_in > 89,
'age_icu_in'] = 91.6
query_output.loc[query_output.age_hosp_in > 89,
'age_hosp_in'] = 91.6
distinct_patients = len(query_output.subject_
id.unique())
hospital_admissions = len(query_output.hadm_
id.unique())
distinct_icu_stay = len(query_output.icustay_
id.unique())
age = query_output.age_icu_in.median()
gender = (query_output.loc[query_output.icu_
seq==1].groupby('gender').gender.count() \
      / query_output.loc[query_output.icu_seq==1].
gender.count()*100)['M']
icu_los = query_output.icu_los.median()
hospital_los = query_output.hosp_los.median()
```

```
icu_mortality = query_output.icu_expire_flag.sum() / float(query_output.icu_expire_flag.count()) * 100
hospital_mortality = (query_output.drop_duplicates(['hadm_id']) \
    .groupby(['hospital_expire_flag']) \
    .hospital_expire_flag.count() \
    / query_output.drop_duplicates(['hadm_id']).hospital_expire_flag.count()*100)[1]
df1 = pd.DataFrame({'Distinct patients': [distinct_patients],
                    'Hospital admissions': [hospital_admissions],
                    'Distinct ICU stays': [distinct_icu_stay],
                    'Age, years, median': [age],
                    'Gender, male, % unit stays': [gender],
                    'ICU length of stay (days)': [icu_los],
                    'Hospital length of stay (days)': [hospital_los],
                    'ICU mortality (%)': [icu_mortality],
                    'Hospital mortality (%)': [hospital_mortality],
                   })
df1
Out[12]:
```

	Age, years, median	Distinct ICU stays	Distinct patients	Gender, male, % unit stays	Hospital admissions	Hospital length of stay (days)	Hospital mortality (%)	ICU length of stay (days)	ICU mortality (%)
0	65.769	53423	38597	56.207693	49785	7.309	11.545646	2.14425	8.545009

Let's now plot some basic demographic information to better visualize the patients

```
In [13]:
died = query_output[query_output.hospital_expire_flag == 1]
```

```
survived = query_output[query_output.icu_expire_
flag == 0]
f, axarr = plt.subplots(2, 2, figsize=(18, 10))
p0 = axarr[0, 0].hist([survived['hosp_los'].drop-
na(), died['hosp_los'].dropna()], stacked=True,
bins=200, normed=False)
axarr[0, 0].set_xlabel('Days in the hospital')
axarr[0, 0].set_ylabel('Number of patients')
axarr[0, 0].set_title('Hospital length of stay')
axarr[0, 0].set_xlim(0, 115)
axarr[0, 0].grid(b=False)
#axarr[0, 0].legend(loc='upper right')
p1 = axarr[0, 1].hist([survived['icu_los'].drop-
na(), died['icu_los'].dropna()], stacked=True,
bins=200, normed=False)
axarr[0, 1].set_xlabel('Days in the ICU')
axarr[0, 1].set_ylabel('Number of patients')
axarr[0, 1].set_title('ICU length of stay')
axarr[0, 1].set_xlim(0, 50)
axarr[0, 1].grid(b=False)
p2=axarr[1, 0].hist([survived['age_hosp_in'].drop-
na(), died['age_hosp_in'].dropna()], stacked=True,
bins=100, normed=False)
axarr[1, 0].set_xlabel('Age (years)')
axarr[1, 0].set_ylabel('Number of patients')
axarr[1, 0].set_title('Age when admitted to the
hospital')
axarr[1, 0].grid(b=False)
p3 = axarr[1, 1].hist([survived['icu_seq'].drop-
na(), died['icu_seq'].dropna()], stacked=True,
bins=25, normed=False)
axarr[1, 1].set_xlabel('Number of times admitted to
the ICU')
axarr[1, 1].set_ylabel('Number of patients')
axarr[1, 1].set_title('ICU Stays')
axarr[1, 1].set_ylim(0, 52000)
axarr[1, 1].set_xlim(0)
```

```
axarr[1, 1].grid(b=False)
axarr[1, 1].legend([‘Survival’, ‘Non-survival’],
loc=(1, 2.3))
plt.tight_layout()
plt.show()
```

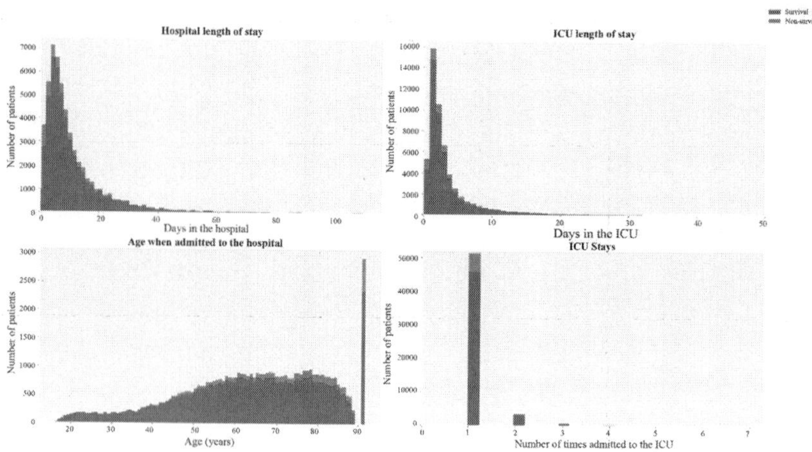

6.2.4. Distinct Patients, no. (% of Total Admissions)

```
In [118]:

print(‘\nTotal number patients: {}.’format(len(query_
output.subject_id.unique())))
print(‘\nNumber of patients by first careunit:\n’)
distinctpatients_df = query_output[[‘first_
careunit’,’subject_id’]] \
                .drop_duplicates([‘subject_id’]) \
                .groupby(‘first_careunit’).count()
print(distinctpatients_df)

print(‘\nProportion of total hospital admissions:\n’)
print(query_output[[‘first_careunit’,’subject_id’]]
\
                .drop_duplicates([‘subject_id’]) \
                    .groupby(‘first_careunit’)
                    .count()/len(query_output.sub-
ject_id.unique())*100)
Total number patients: 38597
```

```
Number of patients by first careunit:
               subject_id
first_careunit
CCU            5674
CSRU           7611
MICU           13649
SICU           6372
TSICU          5291
```

Proportion of total hospital admissions:

```
          subject_id
first_careunit
CCU          14.700624
CSRU         19.719149
MICU         35.362852
SICU         16.509055
TSICU        13.708319
```

6.2.5. Hospital admissions, no. (% of total admissions)

```
In [122]:
print('\nTotal hospital admissions: {}'\
    .format(len(query_output.hadm_id.unique())))
print('\nNumber of hospital admissions by first
careunit:\n')
hospitaladmissions_df    =    query_output[['first_
careunit','hadm_id']] \
                    .drop_duplicates(['hadm_id']) \
                  .groupby('first_careunit').count()

print(hospitaladmissions_df)

print('\nProportion of total hospital admissions:\n')
print(query_output[['first_careunit','hadm_id']] \
                    .drop_duplicates(['hadm_id']) \
                     .groupby('first_careunit') \
                    .count()/len(query_output.hadm_
id.unique())*100)
```

Total hospital admissions: 49785

Number of hospital admissions by first careunit:

```
              hadm_id
first_careunit
CCU             7258
CSRU            8640
MICU           19770
SICU            8110
TSICU           6007
```

Proportion of total hospital admissions:

```
              hadm_id
first_careunit
CCU          14.578688
CSRU         17.354625
MICU         39.710756
SICU         16.290047
TSICU        12.065883
```

6.2.6 Distinct ICU Stays, no. (% of Total Unit Stays)

```
In [11]:
print('\nTotal ICU stays: {}'\
    .format(len(query_output.icustay_id.unique())))
print('\nNumber of ICU stays by careunit:\n')
print(query_output[['first_careunit','icustay_id']]
\
          .groupby('first_careunit').count())
print('\nProportion of total ICU stays:\n')
print(query_output[['first_careunit','icustay_id']]
\
          .groupby('first_careunit') \
                .count()/len(query_output.icustay_
id.unique())*100)
Total ICU stays: 53423
Number of ICU stays by careunit:
```

```
                icustay_id
first_careunit
CCU                   7726
```

```
CSRU          9311
MICU          21087
SICU          8891
TSICU         6408
Proportion of total ICU stays:
       icustay_id
first_careunit
CCU          14.461936
CSRU         17.428823
MICU         39.471763
SICU         16.642645
TSICU        11.994834
```

6.2.7. Age, Yrs, Median ± IQR

We will report the median IQR for the age, since patients aged 90 and older appear with value 300.

```
In [12]:
print('Median age, years: {} .'format(query_output.
age_icu_in.median()))
print('Lower quartile age, years: {} .'format(query_
output.age_icu_in.quantile(0.25)))
print('Upper   quartile   age,   years:   {}   \n
.'format(query_output.age_icu_in.quantile(0.75)))
print('Median age by careunit, years:\n ')
print(query_output[['first_careunit','age_icu_in']]
\
      .groupby('first_careunit').median())
print('\nLower quartile by careunit, years:\n ')
print(query_output[['first_careunit','age_icu_in']]
\
      .groupby('first_careunit').quantile(0.25))
```

```
print('\nUpper quartile by careunit, years:\n ')
print(query_output[['first_careunit','age_icu_in']]
\
      .groupby('first_careunit').quantile(0.75))
```

Median age, years: 65.769

Lower quartile age, years: 52.8361

Upper quartile age, years: 77.80005

Median age by careunit, years:

	age_icu_in
first_careunit	
CCU	70.5697
CSRU	67.9286
MICU	64.9124
SICU	63.5819
TSICU	59.6125

Lower quartile by careunit, years:

	age_icu_in
first_careunit	
CCU	58.44555
CSRU	58.32150
MICU	51.65590
SICU	51.44315
TSICU	42.45665

Upper quartile by careunit, years:

	age_icu_in
first_careunit	
CCU	80.541375
CSRU	76.727250
MICU	78.170950
SICU	76.457500
TSICU	75.601550

6.2.8. Gender, Percent of Unit Stays

```
In [149]:
print('Gender:\n')
print(query_output.loc[query_output.icu_seq==1].
groupby('gender').gender.count())
print(query_output.loc[query_output.icu_seq==1].
groupby('gender').gender.count() \
       /query_output.loc[query_output.icu_seq==1].
gender.count()*100)
print('Gender by careunit:\n')
print(query_output.loc[query_output.icu_seq==1] \
      .groupby(['first_careunit','gender']).gender.
count())
print('\nProportion by unit:\n')
print(query_output.loc[query_output.icu_seq==1] \
    .groupby(['first_careunit','gender']) \
    .gender.count()/query_output.loc[query_output.
icu_seq==1] \
   .groupby(['first_careunit']).gender.count()*100)
Gender:
gender
F    21802
M    27983
Name: gender, dtype: int64
gender
F    43.792307
M    56.207693
```

Name: gender, dtype: float64
Gender by careunit:
first_careunit gender

CCU	F	3055
	M	4203
CSRU	F	2963
	M	5677
MICU	F	9577
	M	10193
SICU	F	3859
	M	4251

```
TSICU        F        2348
             M        3659
Name: gender, dtype: int64
Proportion by unit:
first_careunit  gender
CCU            F       42.091485
               M       57.908515
CSRU           F       34.293981
               M       65.706019
MICU           F       48.442084
               M       51.557916
SICU           F       47.583231
               M       52.416769
TSICU          F       39.087731
               M       60.912269
Name: gender, dtype: float64
```

```
In [ ]:
labels = 'Frogs', 'Hogs', 'Dogs', 'Logs'
sizes = [15, 30, 45, 10]
explode = (0, 0.1, 0, 0)  # only "explode" the 2nd slice (i.e. 'Hogs')
fig1, ax1 = plt.subplots()
ax1.pie(sizes, explode=explode, labels=labels, autopct='%1.1f%%',
        shadow=True, startangle=90)
ax1.axis('equal')  # Equal aspect ratio ensures that pie is drawn as a circle.
plt.show()
```

6.2.9. ICU Length of Stay, Median Days

```
In [14]:
print('Median ICU length of stay, days: {}.'format(query_output.icu_los.median()))
print('Median length of ICU stay by careunit, days:\n ')
print(query_output[['first_careunit','icu_los']] \
      .groupby('first_careunit').median())
```

```
Median ICU length of stay, days: 2.14425
```

Median length of ICU stay by careunit, days:

icu_los

first_careunit

CCU 2.19775

CSRU 2.15295

MICU 2.09560

SICU 2.25220

TSICU 2.11260

6.2.10. ICU Mortality, Percent of Unit Stays

```
In [15]:
print('ICU mortality, number:\n')
print(query_output \
    .groupby(['icu_expire_flag']) \
    .icu_expire_flag.count())
print('\nICU mortality, %:\n')
print(query_output.groupby(['icu_expire_flag']) \
    .icu_expire_flag.count() / query_output.icu_ex-
pire_flag.count()*100)
print('\nICU mortality by careunit:\n')
print(query_output \
    .groupby(['first_careunit','icu_expire_flag']) \
    .icu_expire_flag.count())
print('\nProportion by unit:\n')
print(query_output \
    .groupby(['first_careunit','icu_expire_flag']) \
    .icu_expire_flag.count()/query_output \
       .groupby(['first_careunit']).icu_expire_flag.
count()*100)
```

ICU mortality, number:

icu_expire_flag

0 48858

1 4565

Name: icu_expire_flag, dtype: int64

```
ICU mortality, %:
icu_expire_flag
0    91.454991
1     8.545009
Name: icu_expire_flag, dtype: float64
ICU mortality by careunit:
first_careunit  icu_expire_flag
CCU             0                  7041
                1                   685
CSRU            0                  9010
                1                   301
MICU            0                 18865
                1                  2222
SICU            0                  8078
                1                   813
TSICU           0                  5864
                1                   544
Name: icu_expire_flag, dtype: int64
Proportion by unit:
first_careunit  icu_expire_flag
CCU             0                  91.133834
                1                   8.866166
CSRU            0                  96.767265
                1                   3.232735
MICU            0                  89.462702
                1                  10.537298
SICU            0                  90.855922
                1                   9.144078
TSICU           0                  91.510612
                1                   8.489388
Name: icu_expire_flag, dtype: float64
```

6.2.11. Hospital Mortality, Percent Unit of Stays

```
In [165]:
# NB: hadm_id is repeated in data due to multiple
ICU stays
# ...so need to drop duplicates
print('Hospital mortality, number:\n')
print(query_output.drop_duplicates(['hadm_id']) \
    .groupby(['hospital_expire_flag']) \
    .hospital_expire_flag.count())
print('\nHospital mortality, %:\n')
print(query_output.drop_duplicates(['hadm_id']) \
    .groupby(['hospital_expire_flag']) \
    .hospital_expire_flag.count() \
      / query_output.drop_duplicates(['hadm_id']).
hospital_expire_flag.count()*10)
print('\nHospital mortality:\n')
print(query_output.drop_duplicates(['hadm_id']) \
        .groupby(['first_careunit','hospital_expire_
flag']) \
    .hospital_expire_flag.count())
print('\nProportion by unit:\n')
print(query_output.drop_duplicates(['hadm_id']) \
        .groupby(['first_careunit','hospital_expire_
flag']) \
   .hospital_expire_flag.count()/query_output.drop_
duplicates(['hadm_id']) \
       .groupby(['first_careunit']).hospital_expire_
flag.count()*100)
Hospital mortality, number:
hospital_expire_flag
```

0 44037

1 5748

Name: hospital_expire_flag, dtype: int64

Hospital mortality, %:

hospital_expire_flag

0 88.454354

1 11.545646

```
Name: hospital_expire_flag, dtype: float64
Hospital mortality:
first_careunit  hospital_expire_flag
CCU             0                       6441
                1                       817
CSRU            0                       8277
                1                       363
MICU            0                       16911
                1                       2859
SICU            0                       7090
                1                       1020
TSICU           0                       5318
                1                       689
Name: hospital_expire_flag, dtype: int64
Proportion by unit:
first_careunit  hospital_expire_flag
CCU             0                       88.743455
                1                       11.256545
CSRU            0                       95.798611
                1                       4.201389
MICU            0                       85.538695
                1                       14.461305
SICU            0                       87.422935
                1                       12.577065
TSICU           0                       88.530048
                1                       11.469952
Name: hospital_expire_flag, dtype: float64
```

6.2.12. Patient Characteristics

Let's first see how many charted observations are available for each hospitalization. To do so, we'll use the CHARTEVENTS table which contains the charted data, or the electronic chart of a patient. The electronic chart holds a patient's routine vital signs and additional information relevant to their care. The entries contain the SUBJECT_ID, HADM_ID and ICUSTAY_ID identified, and are associated with an ITEMID which is an identifier for a single measurement type, a CHARTTIME which is the timestamp at which the observation was made and a VALUE which contains the value measured identified by ITEMID. The ITEMID identifier description can be found in the D_ITEMS table which holds additional information regarding that measurement type, e.g., ITEMID 212 corresponds to a heart rate measurement.

```
In [31]:
query = \
"""
WITH chartobs AS (
SELECT hadm_id, COUNT(DISTINCT itemid) as variable,
count(hadm_id) as obs
FROM chartevents
GROUP BY hadm_id)
SELECT avg(obs) as avg_obs, avg(variable) as avg_
variable
FROM chartobs;
"""
query_output = pd.read_sql_query(query,con)
query_output.head()
Out[31]:
```

	avg_obs	avg_variable
0	5774.418267	234.783262

```
In [27]:
query = \
"""
SELECT *
FROM D_ITEMS;
"""
items_df = pd.read_sql_query(query,con)
print("Number of different measurement types:",
len(items_df))
items_df.head()
```

```
('Number of different measurement types:', 12487)
Out[27]:
```

	row_id	itemid	label	abbreviation	dbsource	linksto	category	unitname	param_type	conceptid
0	457	497	Patient controlled analgesia (PCA) [Inject]	None	carevue	chartevents	None	None	None	None
1	458	498	PCA Lockout (Min)	None	carevue	chartevents	None	None	None	None
2	459	499	PCA Medication	None	carevue	chartevents	None	None	None	None
3	460	500	PCA Total Dose	None	carevue	chartevents	None	None	None	None
4	461	501	PCV Exh Vt (Obser)	None	carevue	chartevents	None	None	None	None

As seen from above, the database contains a huge amount of data (over 12,000 different measurement types), with each hospitalization having on average over 230 different types of measurements and over 5700-charted observations. For that reason, we'll have to carefully select the columns and measurement types from the patient data in order to do our analysis.

6.2.13. Data for a Single Patient

Now that we've seen a description of the overall population, let's investigate data from a single patient. We'll use the ADMISSIONS, PATIENTS and ICUSTAYS tables. We'll focus on a single ICU stay for a patient and extract the chart events for that identifier using the CHARTEVENT table and the OUTPUTEVENTS which is the output information for patients while in the ICU. We'll then have to link these tables with dictionary tables that provide definitions for the identifiers. For instance, every row in OUTPUT events is associated with a single ITEMID which represents the type of measurement. The actual name of the drug is found by joining OUTPUTEVENTS and D_ITEMS on ITEMID.

```
In [33]:
# Select a single ICU stay
icustay_id = 293325
# Load admissions
query = """
SELECT ad.admission_type
```

```
  , ad.diagnosis
  , ic.icustay_id
  , ic.first_careunit
  , ic.last_careunit
  , ic.intime as icu_intime
  , ad.hospital_expire_flag
  , pa.expire_flag
FROM admissions ad
INNER JOIN icustays ic
ON ad.subject_id = ic.subject_id
INNER JOIN patients pa
ON ad.subject_id = pa.subject_id
WHERE ic.icustay_id = "'" + str(icustay_id) + "'"
ORDER BY ic.intime
"""
admissions = pd.read_sql_query(query,con)
admissions.head()
Out[33]:
```

	admission_type	diagnosis	icustay_id	first_careunit	last_ca-reunit	icu_intime	hospital_expire_flag	expire_flag
0	EMERGENCY	NEUTROPENIC FEVER	293325	MICU	MICU	2120-10-29 10:32:50	1	1

Let's start with chartevents, which are vital signs. Vital signs are clinical measurements that describe the state of a patient's body functions. The queried data comes with an icustay_id identifying what ICU stay this measurement is associated with, along with a timestamp, a label indicating what type of measurement it is and a value.

```
In [34]:
# Load chartevents
query = """
SELECT ce.icustay_id
  , ce.charttime
  , ce.charttime - ie.intime AS icutime
  , di.label
  , ce.value
  , ce.valuenum
```

```
  , ce.valueuom
FROM chartevents ce
INNER join d_items di
ON ce.itemid = di.itemid
INNER join icustays ie
ON ce.icustay_id = ie.icustay_id
WHERE ce.icustay_id = """ + str(icustay_id) + """
AND error != 1
ORDER BY ce.charttime
"""
charts = pd.read_sql_query(query,con)
charts.head()
Out[34]:
```

	icustay_id	charttime	icutime	label	value	valuenum	valueuom
0	293325	2120-10-29 10:30:00	-1 days +23:57:10	Inspired O2 Fraction	70	70.0	None
1	293325	2120-10-29 10:30:00	-1 days +23:57:10	O2 Delivery Device(s)	Aerosol-cool	NaN	None
2	293325	2120-10-29 10:30:00	-1 days +23:57:10	O2 Flow	10	10.0	L/min
3	293325	2120-10-29 10:32:00	-1 days +23:59:10	Marital Status	Single	1.0	None
4	293325	2120-10-29 10:32:00	-1 days +23:59:10	Language	English	1.0	None

```
In [37]:
# Load outputevents
query = """
SELECT oe.icustay_id
  , oe.charttime
  , oe.charttime - ie.intime AS icutime
  , di.label
  , oe.value
  , oe.valueuom
FROM outputevents oe
INNER JOIN icustays ie
  ON oe.icustay_id = ie.icustay_id
INNER JOIN d_items di
  ON oe.itemid = di.itemid
WHERE oe.icustay_id = """ + str(icustay_id) + """
ORDER BY oe.charttime
"""
```

```
outputs = pd.read_sql_query(query,con)
outputs.head()
Out[37]:
```

	icustay_id	charttime	icutime	label	value	valueuom
0	293325	2120-10-29 10:45:00	00:12:10	Foley	50.0	mL
1	293325	2120-10-29 11:30:00	00:57:10	Foley	100.0	mL
2	293325	2120-10-29 12:00:00	01:27:10	Foley	35.0	mL
3	293325	2120-10-29 13:00:00	02:27:10	Foley	50.0	mL
4	293325	2120-10-29 14:00:00	03:27:10	Foley	40.0	mL

Laboratory measurements are made by acquiring a fluid from the patient's body (e.g., blood from an arterial line or urine from a catheter) and then analyzing it in the laboratory.

```
In [39]:
# Load labevents
query = """
SELECT ie.icustay_id
  , le.charttime
  , le.charttime - ie.intime AS icutime
  , di.label
  , le.value
  , le.valuenum
  , le.valueuom
FROM labevents le
INNER join icustays ie
  ON le.subject_id = ie.subject_id
  AND le.charttime >= ie.intime
  AND le.charttime <= ie.outtime
INNER JOIN d_labitems di
  ON le.itemid = di.itemid
WHERE ie.icustay_id = """ + str(icustay_id) + """
ORDER BY le.charttime
"""
labs = pd.read_sql_query(query,con)
labs.head()
Out[39]:
```

	icustay_id	charttime	icutime	label	value	valuenum	valueuom
0	293325	2120-10-29 11:42:00	01:09:10	Red Blood Cells	2.64	2.64	m/uL
1	293325	2120-10-29 11:42:00	01:09:10	Lactate Dehydrogenase (LD)	1321	1321.00	IU/L
2	293325	2120-10-29 11:42:00	01:09:10	Magnesium	1.6	1.60	mg/dL
3	293325	2120-10-29 11:42:00	01:09:10	Basophils	0	0.00	%
4	293325	2120-10-29 11:42:00	01:09:10	Bands	35	35.00	%

We can now determine which are the most commonly measured variables in charts and lab events and the average values for physiological signs of the patient.

```
In [40]:
# Create column with minutes from ICU intime
charts['icutimehr'] = (charts['icutime'].dt.second
s/60/60)+(charts['icutime'].dt.days*24)
outputs['icutimehr'] = (outputs['icutime'].dt.seco
nds/60/60)+(outputs['icutime'].dt.days*24)
labs['icutimehr'] = (labs['icutime'].dt.seconds/60
/60)+(labs['icutime'].dt.days*24)
# Snip data to specified number of days
maxdays = 5;
charts=charts.loc[charts.icutime.dt.days<=maxdays]
outputs = outputs.loc[outputs.icutime.
dt.days<=maxdays]
labs = labs.loc[labs.icutime.dt.days<=maxdays]
In [41]:
print(charts['label'].value_counts().head())
print()
print(labs.mean())
print()
# Get average values
hr_mean = charts.valuenum[charts.label=='Heart
Rate'].mean()
bp_mean = charts.icutimehr[charts.label=='Non In-
vasive Blood Pressure mean'].mean()
temp_mean = ((charts.valuenum[charts.
label=='Temperature Fahrenheit']-32)/1.8).mean()
# Print averages
print("Mean HR is: {0:.2f}."format(hr_mean))
```

```
print("Mean BP is: {0:.2f}."format(bp_mean))
print("Mean temp, C is: {0:.2f}."format(temp_mean))
```

```
    O2 saturation pulseoxymetry    62
    Respiratory Rate               62
    Heart Rate                     61
    Heart Rhythm                   57
    Ectopy Type 1                  56
    Name: label, dtype: int64
()
icustay_id                 293325
icutime      0 days 13:15:47.440000
valuenum                  57.8909
icutimehr                 13.2632
dtype: object
()
Mean HR is: 98.89
Mean BP is: 22.62
Mean temp, C is: 36.33
```

6.2.14. Plotting Sample Data Over First 24 Hours From Admission to ICU

In [197]:

```
# Credit: Randal Olson for styling (http://www.randalolson.com/2014/06/28/)
# Prepare the size of the figure
fig = plt.figure(figsize=(22, 20))
plt.rcParams.update({'font.size': 22})
# "Tableau 20" colors as RGB.
tableau20 = [(31, 119, 180), (174, 199, 232), (255, 127, 14), (255, 187, 120),
        (44, 160, 44), (152, 223, 138), (214, 39, 40), (255, 152, 150),
        (148, 103, 189), (197, 176, 213), (140, 86, 75), (196, 156, 148),
        (227, 119, 194), (247, 182, 210), (127, 127, 127), (199, 199, 199),
        (188, 189, 34), (219, 219, 141), (23, 190, 207), (158, 218, 229)]
```

```
# Scale the RGB values to the [0, 1] range, which
is the format matplotlib accepts.
for i in range(len(tableau20)):
    r, g, b = tableau20[i]
    tableau20[i] = (r / 255., g / 255., b / 255.)

# Remove the plot frame lines.
ax = plt.subplot(111)
ax.spines["top"].set_visible(False)
ax.spines["bottom"].set_visible(True)
ax.spines["right"].set_visible(False)
ax.spines["left"].set_visible(True)

# Ensure that the axis ticks only show up on the
bottom and left of the plot.
ax.get_xaxis().tick_bottom()
ax.get_yaxis().tick_left()
ax.axis([0,60,0,150])
# Plot vital signs
plt.plot(charts.icutimehr[charts.label=='Heart
Rate'],
        charts.valuenum[charts.label=='Heart Rate'],
          color=tableau20[6], lw=2.5,
            marker='o', markersize=6, label='Heart
rate')
plt.plot(charts.icutimehr[charts.label=='O2  satu-
ration pulseoxymetry'],
          charts.valuenum[charts.label=='O2 satura-
tion pulseoxymetry'],
          color=tableau20[1], lw=2.5,
          marker='o', markersize=6, label='O2 satu-
ration')
plt.plot(charts.icutimehr[charts.label=='Non Inva-
sive Blood Pressure mean'],
        charts.valuenum[charts.label=='Non Invasive
Blood Pressure mean'],
          color=tableau20[4], lw=2.5,
            marker='o', markersize=6, label='NIBP,
mean')
```

```
plt.plot(charts.icutimehr[charts.
label=='Respiratory Rate'],
         charts.valuenum[charts.label=='Respiratory
Rate'],
          color=tableau20[2], lw=2.5,
        marker='o', markersize=6, label='Respiratory
rate')

plt.legend(loc=5,fontsize=18)
plt.xlabel('Time after admission to the intensive
care unit, hours', fontsize=22)
plt.ylabel('Measurement, absolute value', font-
size=22)
plt.yticks(np.arange(0, 140, 20))
Out[197]:
([<matplotlib.axis.YTick at 0x7f192db2e438>,
  <matplotlib.axis.YTick at 0x7f192d816048>,
  <matplotlib.axis.YTick at 0x7f192f914eb8>,
  <matplotlib.axis.YTick at 0x7f192d0f26d8>,
  <matplotlib.axis.YTick at 0x7f192e50fbe0>,
  <matplotlib.axis.YTick at 0x7f192d0e04a8>,
  <matplotlib.axis.YTick at 0x7f192d0e0eb8>],
 <a list of 7 Text yticklabel objects>)
```

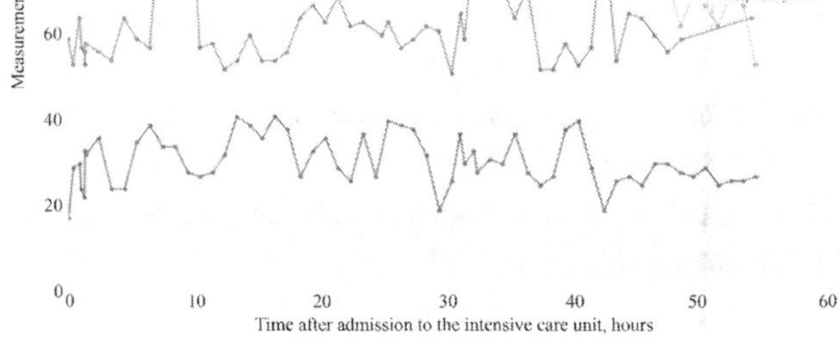

```
In [11]:
query = \
"""
with ce as
(
 select adm.hadm_id, adm.subject_id, chart.icustay_id
 , chart.charttime
 , chart.charttime – ie.intime AS icutime

 --, min(case when itemid in (211,220045) and valuenum > 0 then valuenum else null end) as HeartRate_Min
 --, max(case when itemid in (211,220045) and valuenum < 300 then valuenum else null end) as HeartRate_Max
  , case when itemid in (211,220045) and valuenum > 0 then valuenum else null end as heartrate
 from admissions adm
 left join chartevents chart
  on adm.hadm_id = chart.hadm_id
  and chart.itemid in
  (
  -- HEART RATE
  211, --"Heart Rate"
  220045 --"Heart Rate"
  )
 left join icustays ie
 -- match the tables on the patient identifier
 on ie.icustay_id = chart.icustay_id
 -- and require that the observation be made after the patient is admitted to the ICU
 and chart.charttime >= ie.intime
 -- and *before* their admission time + 1 day, i.e. the observation must be made on their first day in the ICU
 and chart.charttime <= ie.intime + interval '1 day'
 and chart.error IS DISTINCT FROM 1
  group by adm.hadm_id, chart.icustay_id, adm.subject_id, chart.charttime, ie.intime, chart.itemid, valuenum
)
select *
from ce
```

```
where ce.subject_id = 3;
"""
vitals = pd.read_sql_query(query, con)
vitals.head()
Out[11]:
```

	hadm_id	subject_id	icustay_id	charttime	icutime	heartrate
0	145834	3	211552	2101-10-20 18:45:00	NaT	95.0
1	145834	3	211552	2101-10-20 19:30:00	00:19:49	151.0
2	145834	3	211552	2101-10-20 19:45:00	00:34:49	135.0
3	145834	3	211552	2101-10-20 20:00:00	00:49:49	143.0
4	145834	3	211552	2101-10-20 20:15:00	01:04:49	165.0

```
In [14]:
min(vitals.heartrate)
Out[14]:
64.0
```

6.3. VISUALIZATIONS

```
In [2]:
query_output = pd.read_csv('./full-features-materialized.csv') #.dropna().reset_index(drop=True)
query_output.first_careunit = pd.Categorical(query_output.first_careunit)
query_output.gender = pd.Categorical(query_output.gender)
query_output.admission_type = pd.Categorical(query_output.admission_type)
query_output['gender'] = query_output.gender.cat.codes
query_output['first_careunit'] = query_output.first_careunit.cat.codes
query_output['admission_type'] = query_output.admission_type.cat.codes
query_output = query_output[(query_output.exclu-
```

```
sion_los == 0) & (query_output.exclusion_age == 0)
                                   & (query_output.exclu-
sion_first_stay == 0) ]#& (query_output.age_icu_in
< 250)]
# Median age of patients > 89 is 91.6
query_output.loc[query_output.age_icu_in   >   89,
'age_icu_in'] = 91.6
mortality_names = ['hospital_expire_flag', 'icu_ex-
pire_flag',
                  'hospital1year_expire_flag', 'hos-
pital30day_expire_flag']
los_names = ['hosp_los', 'icu_los']
feature_names = [
            # Demographic information
           'age_icu_in', 'gender', 'first_careunit',
            'admission_type',

            # Vital signs
            'heartrate_min', 'heartrate_max',
            'meanbp_min', 'meanbp_max',
            'diasbp_min', 'diasbp_max',
            'sysbp_min', 'sysbp_max',
            'resprate_min', 'resprate_max',
            'mingcs', 'gcsmotor',
            'gcsverbal', 'gcseyes',
            'tempc_min', 'tempc_max',
            'spo2_min', 'spo2_max',

            # Laboratory measurements

            'aniongap_min', 'aniongap_max',
            #'albumin_min', 'albumin_max',
            #'bands_min', 'bands_max',
            'bicarbonate_min', 'bicarbonate_max',
            #'bilirubin_min', 'bilirubin_max',
            'creatinine_min', 'creatinine_max',
            'chloride_min', 'chloride_max',
            'glucose_min', 'glucose_max',
            'hematocrit_min', 'hematocrit_max',
```

```
            'hemoglobin_min', 'hemoglobin_max',
            #'lactate_min', 'lactate_max',
            'platelet_min', 'platelet_max',
            'potassium_min', 'potassium_max',
            #'ptt_min', 'ptt_max',
            #'inr_min', 'inr_max',
            #'pt_min', 'pt_max',
            'sodium_min', 'sodium_max',
            'bun_min', 'bun_max',
            'wbc_min', 'wbc_max',

            # Other
            'vent',
            'urineoutput'
            ]
query_output = query_output.loc[:, mortality_names
+ feature_names]
query_output = query_output.fillna(query_output.
mean()).reset_index(drop=True)
#query_output = query_output.loc[:, mortality_names
+ feature_names].dropna().reset_index(drop=True) #
drop NaNs
outcomes = query_output.loc[:, mortality_names]
features = query_output.loc[:, query_output.col-
umns.difference(mortality_names + los_names)]
X = features
y = outcomes['hospital_expire_flag'].replace(0, -1)
print(len(query_output))
query_output.tail()
32290
Out[2]:
```

	hospital_expire_flag	icu_expire_flag	hospital1year_expire_flag	hospital30day_expire_flag	age_icu_in	gender	first_careunit	admission_type	heartrate_min	heartrate_max	...	potassium_min	potassium_max	sodium_min	sodium_max	bun_min	bun_max	wbc_min	wbc_max	vent	urineoutput
32285	0	0	0	0	70.826046	0	5	0	62.0	119.0	...	3.1	3.9	135.0	144.0	7.0	7.0	3.3	4.5	1	1995.0
32286	0	0	0	0	63.893494	0	2	1	60.0	95.0	...	3.5	3.9	139.0	143.0	16.0	31.0	5.3	8.6	0	1400.0
32287	0	0	0	0	43.594414	0	0	1	58.0	80.0	...	3.8	4.1	139.0	140.0	12.0	13.0	11.7	11.8	0	1250.0
32288	1	1	1	1	86.957864	1	2	1	52.0	68.0	...	5.0	5.7	123.0	128.0	41.0	42.0	11.4	13.3	1	730.0
32289	0	0	0	0	77.488270	1	0	1	28.0	74.0	...	4.4	4.9	135.0	138.0	21.0	22.0	10.8	11.0	0	1800.0

5 rows × 52 columns

6.3.1. 3D Plot with 3 Features

Although we now have a lot of features extracted, let's inspect how they might have an impact on mortality prediction. We'll use a 3D scatter plot to visualize age, heart rate and the Glasgow Coma Scale of patients.

```
In [3]:
from mpl_toolkits.mplot3d import Axes3D
non_survival = query_output.loc[query_output.hos-
pital_expire_flag == 1, ['age_icu_in', 'heartrate_
min', 'mingcs', ]]
survival = query_output.loc[query_output.hospital_
expire_flag == 0, ['age_icu_in', 'heartrate_min',
'mingcs', ]]
fig = plt.figure(figsize=(18, 12))
ax = fig.add_subplot(111, projection='3d')
ax.scatter(non_survival.loc[:, non_survival.col-
umns[0]], non_survival.loc[:, non_survival.col-
umns[1]],
              non_survival.loc[:, non_survival.col-
umns[2]], c='red', marker='x', cmap='coolwarm',
label='Died in hospital')
ax.scatter(survival.loc[:, survival.columns[0]],
survival.loc[:, survival.columns[1]],
     survival.loc[:, survival.columns[2]], c='blue',
marker='+', cmap='coolwarm', label='Survived')
ax.set_xlabel('Age', fontsize=16)
ax.set_ylabel('Minimum Heartrate', fontsize=16)
ax.set_zlabel('Glasgow Coma Scale', fontsize=16)
ax.legend(loc='upper right', fontsize=18)
ax.set_title('3D Plot of 3 predictors for mortality
prediction', fontsize=20)
plt.show()
```

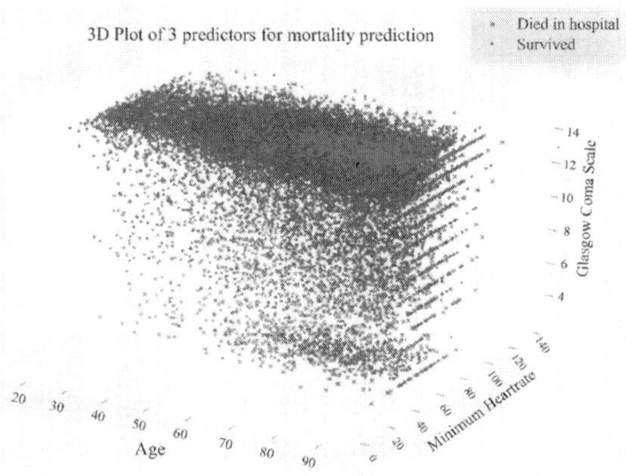

6.3.2. First Lab Measurement Visualization

To get a better idea of how laboratory measurements for patients admitted to the ICU, we'll plot the distribution of measurements for survival and non-survival groups for various lab measurements. Normal range for the lab measurements can be seen below which have a lower and upper limit.

Lab Value	Lower limit	Upper Limit	Units
Bicarbonate	22	32	mEq/L
BUN	6	20	mEq/L
Calcium	8.4	10.3	mg/dL
Chloride	96	108	mEq/L
Creatinine	0.4	1.1	mEq/L
Hemoglobin	11.2	15.7	g/dL
Lactate	0.5	2	mmol/L
Magnesium	1.6	2.6	mg/dL
Phosphate	2.7	4.5	mg/dL
Platelet count	150	400	K/uL
Potassium	3.3	5.1	mEq/L
Sodium	133	145	mEq/L

```
In [11]:
# Create dictionary of ranges
# Keys should match the lab names in the query below
# Are these values correct given the units? Need to
check.
lab_ranges = {'BICARBONATE': [22,32],
              'BUN': [6,20],
```

```
               'CALCIUM': [8.4,10.3],
               'CHLORIDE': [96,108],
               'CREATININE': [0.4,1.1],
               'HEMOGLOBIN': [11.2,15.7],
               'LACTATE': [0.5,2.0],
               'PHOSPHATE': [2.7,4.5],
               'PLATELET': [150,400],
               'POTASSIUM': [3.3,5.1],
               'SODIUM': [133,145]
              }
labs = {'ANIONGAP': 'mEq/L',
        #'ALBUMIN': 'g/dL',
        'BICARBONATE': 'mEq/L',
        #'BILIRUBIN': 'mg/dL',
        'BUN': '',
        'CHLORIDE': 'mEq/L',
        'CREATININE': 'mg/dL',
        'GLUCOSE': 'mg/dL',
        'HEMATOCRIT': '%',
        'HEMOGLOBIN': 'g/dL',
        #'INR': '',
        #'LACTATE': 'mmol/L',
        'PLATELET': 'K/uL',
        'POTASSIUM': 'mEq/L',
        #'PT': '',
        #'PTT': 'sec',
        'SODIUM':'mmol/L',
        'WBC': ''}
data = query_output
for l, u in iter(sorted(labs.iteritems())):
    # count patients
   n_nonsurv = data[l.lower()+'_max'][data.icu_ex-
pire_flag==1].dropna().count()
     n_surv  = data[l.lower()+'_max'][data.icu_ex-
pire_flag==0].dropna().count()

    # get median, variance, skewness
    med_nonsurv = data[l.lower()+'_max'][data.icu_
expire_flag==1].dropna().median()
```

```
    med_surv = data[l.lower()+'_max'][data.icu_ex-
pire_flag==0].dropna().median()
    var_nonsurv = data[l.lower()+'_max'][data.icu_
expire_flag==1].dropna().var()
    var_surv = data[l.lower()+'_max'][data.icu_ex-
pire_flag==0].dropna().var()
      skew_nonsurv = data[l.lower()+'_max'][data.
icu_expire_flag==1].dropna().skew()
    skew_surv = data[l.lower()+'_max'][data.icu_ex-
pire_flag==0].dropna().skew()

    # plot
    plt.subplots(figsize=(13,6))
   data[l.lower()+'_max'][data.icu_expire_flag==1].
dropna().plot.kde(
    alpha=1.0,label='Non-survival (n={}).'format(n_
nonsurv))
   data[l.lower()+'_max'][data.icu_expire_flag==0].
dropna().plot.kde(
        alpha=1.0,label='Survival (n={}).'format(n_
surv))

    # fake plots for KS test, median, etc
    plt.plot([], label=' ',color='lightgray')
          plt.plot([],  label='Median  (non-surv):
{}.'format(format(med_nonsurv,.'2f')),
             color='lightgray')
            plt.plot([],   label='Median   (surv):
{}.'format(format(med_surv,.'2f')),
             color='lightgray')
        plt.plot([],  label='Variance  (non-surv):
{}.'format(format(var_nonsurv,.'2f')),
             color='lightgray')
          plt.plot([],   label='Variance   (surv):
{}.'format(format(var_surv,.'2f')),
             color='lightgray')
          plt.plot([],   label='Skew   (non-surv):
{}.'format(format(skew_nonsurv,.'2f')),
             color='lightgray')
```

```
            plt.plot([],   label='Skew   (surv):
{}.'format(format(skew_surv,.'2f')),
             color='lightgray')
    # add title, labels etc.
   plt.title('First {} measurement on ICU admission
.'format(l.lower()) +
               'vs ICU mortality \n')
    plt.xlabel(l + ' ' + u)
          plt.legend(loc="upper  left",  bbox_to_
anchor=(1,1),fontsize=12)
         plt.xlim(0,  data[l.lower()+'_max'].quan-
tile(0.99))

    # Add lab range if available
    if l in lab_ranges:
                          plt.axvline(lab_ranges[l]
[0],color='k',linestyle='--')
                          plt.axvline(lab_ranges[l]
[1],color='k',linestyle='--')
```

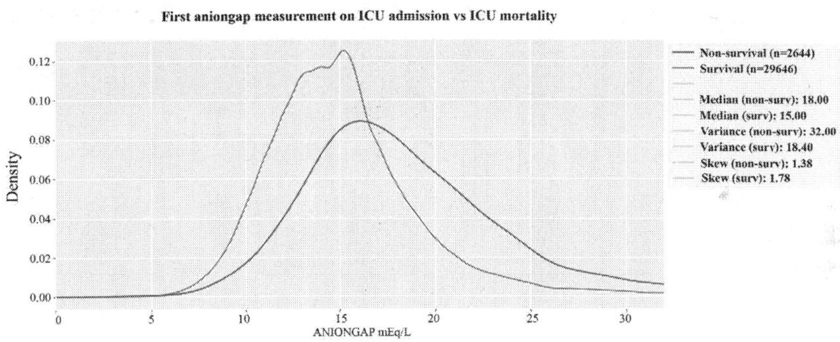

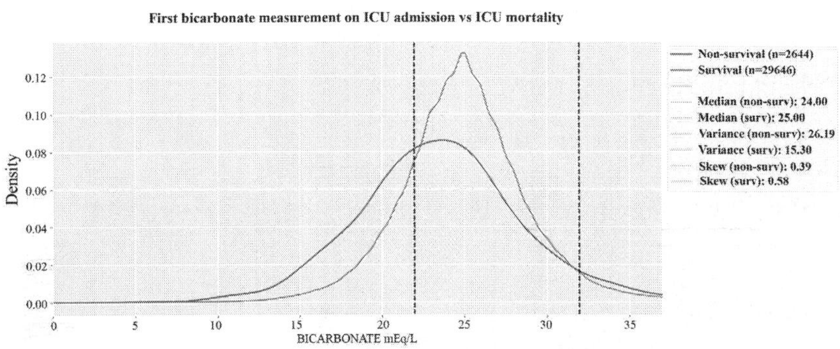

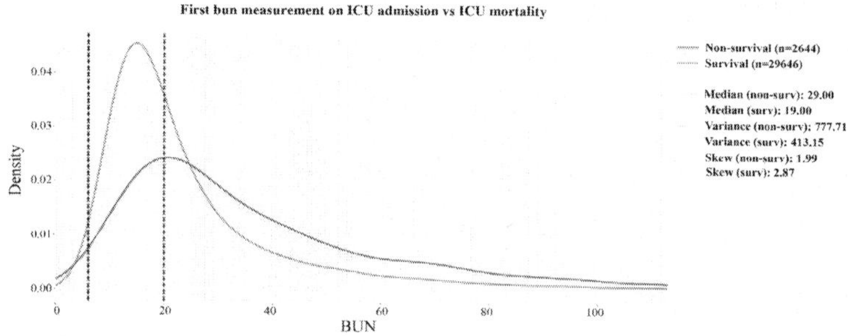
First bun measurement on ICU admission vs ICU mortality
Non-survival (n=2644)
Survival (n=29646)
Median (non-surv): 29.00
Median (surv): 19.00
Variance (non-surv): 777.71
Variance (surv): 413.15
Skew (non-surv): 1.99
Skew (surv): 2.87
Density
BUN

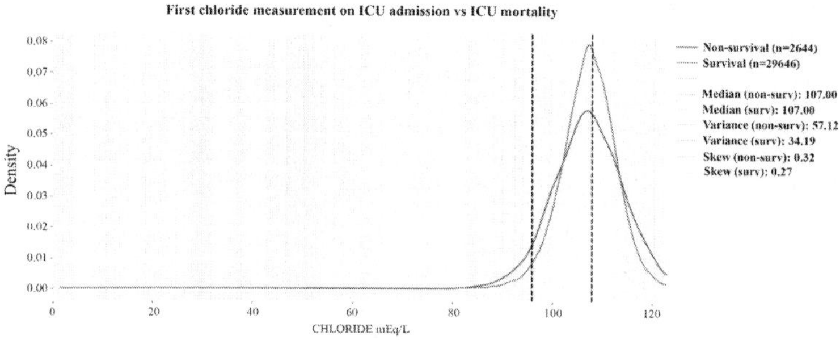
First chloride measurement on ICU admission vs ICU mortality
Non-survival (n=2644)
Survival (n=29646)
Median (non-surv): 107.00
Median (surv): 107.00
Variance (non-surv): 57.12
Variance (surv): 34.19
Skew (non-surv): 0.32
Skew (surv): 0.27
Density
CHLORIDE mEq/L

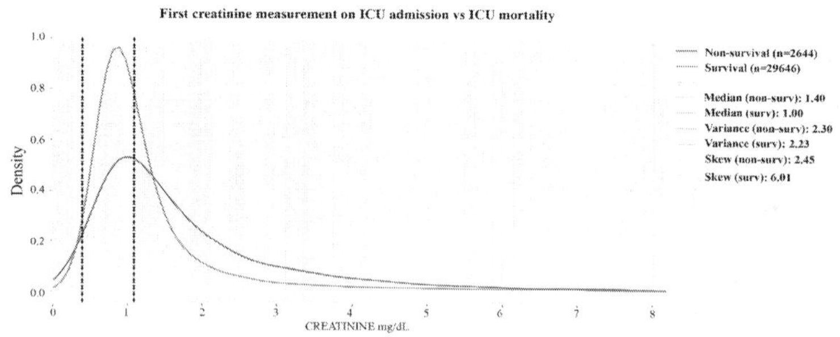
First creatinine measurement on ICU admission vs ICU mortality
Non-survival (n=2644)
Survival (n=29646)
Median (non-surv): 1.40
Median (surv): 1.00
Variance (non-surv): 2.30
Variance (surv): 2.23
Skew (non-surv): 2.45
Skew (surv): 6.01
Density
CREATININE mg/dL

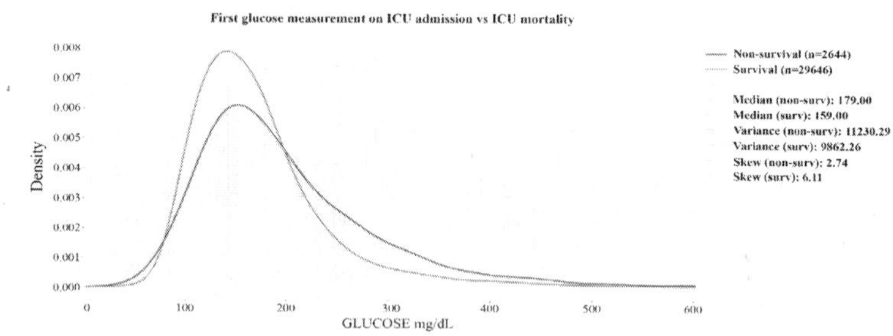
First glucose measurement on ICU admission vs ICU mortality
Non-survival (n=2644)
Survival (n=29646)
Median (non-surv): 179.00
Median (surv): 159.00
Variance (non-surv): 11230.29
Variance (surv): 9862.26
Skew (non-surv): 2.74
Skew (surv): 6.11
Density
GLUCOSE mg/dL

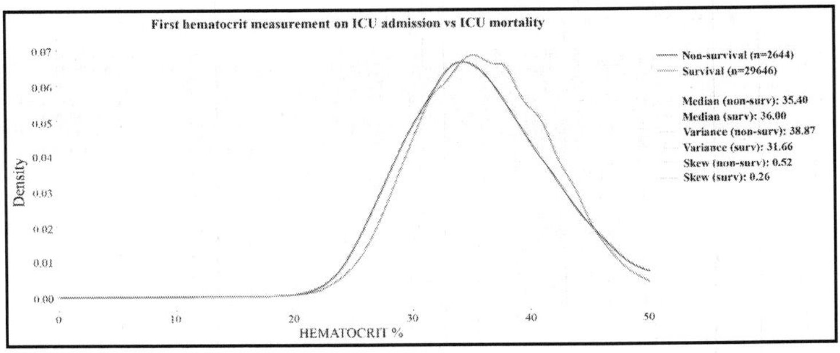
First hematocrit measurement on ICU admission vs ICU mortality
Non-survival (n=2644)
Survival (n=29646)
Median (non-surv): 35.40
Median (surv): 36.00
Variance (non-surv): 38.87
Variance (surv): 31.66
Skew (non-surv): 0.52
Skew (surv): 0.26
Density
HEMATOCRIT %

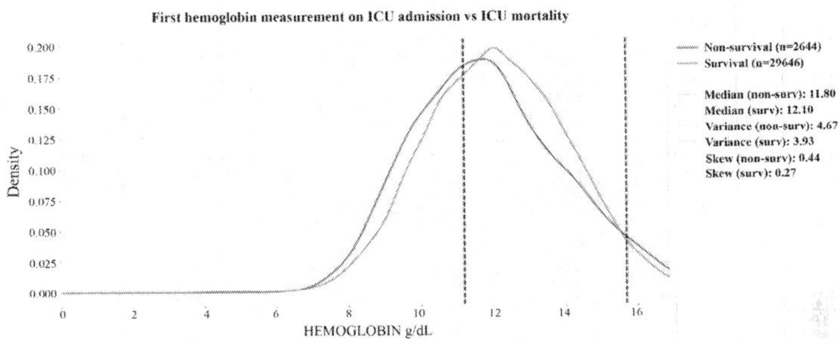
First hemoglobin measurement on ICU admission vs ICU mortality
Non-survival (n=2644)
Survival (n=29646)
Median (non-surv): 11.80
Median (surv): 12.10
Variance (non-surv): 4.67
Variance (surv): 3.93
Skew (non-surv): 0.44
Skew (surv): 0.27
Density
HEMOGLOBIN g/dL

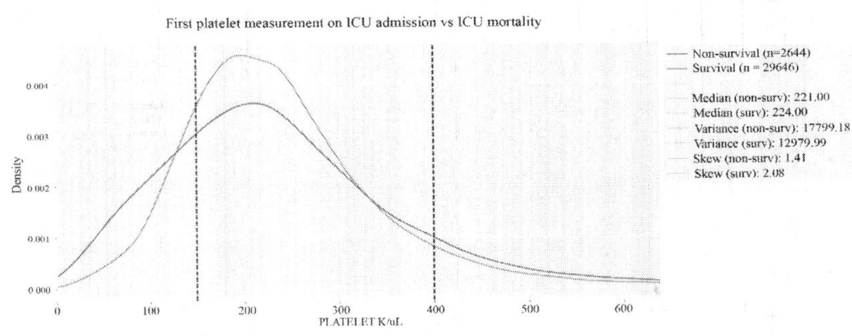
First platelet measurement on ICU admission vs ICU mortality
Non-survival (n=2644)
Survival (n = 29646)
Median (non-surv): 221.00
Median (surv): 224.00
Variance (non-surv): 17799.18
Variance (surv): 12979.99
Skew (non-surv): 1.41
Skew (surv): 2.08
Density
PLATELET K/uL

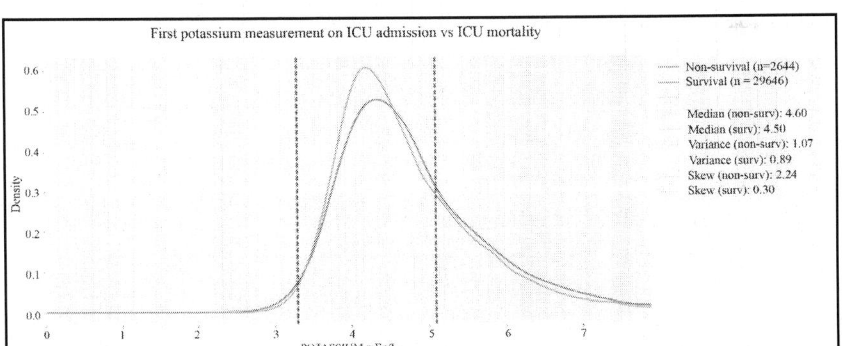
First potassium measurement on ICU admission vs ICU mortality
Non-survival (n=2644)
Survival (n = 29646)
Median (non-surv): 4.60
Median (surv): 4.50
Variance (non-surv): 1.07
Variance (surv): 0.89
Skew (non-surv): 2.24
Skew (surv): 0.30
Density
POTASSIUM mEq/L

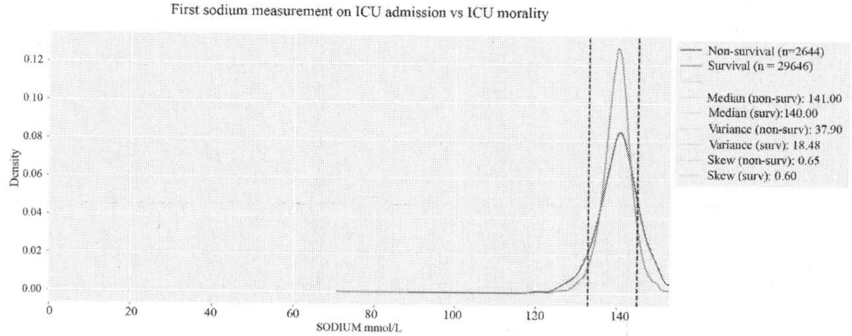

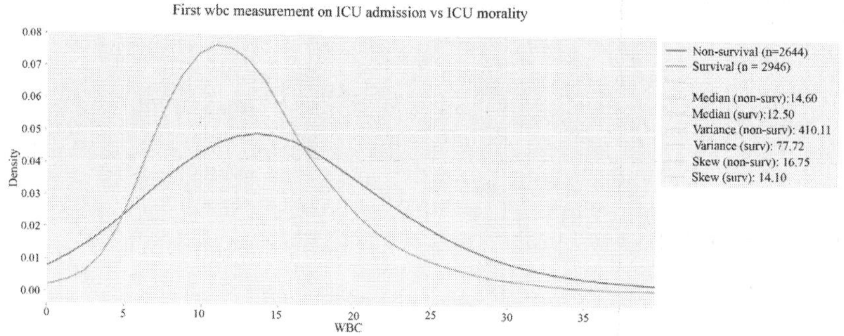

6.4. FEATURE SELECTION AND DIMENSIONALITY REDUCTION

I extracted predictors from three main categories: demographic information, vital sign data and laboratory measurements. These were selected as the most relevant information in determining the likelihood of mortality and hospital length of stay.

Demographic & Clinical Info	Description
Age	Age of the patient upon entering the ICU
Gender	Patient gender (male or female)
Hospital length of stay	Number of days spent in the hospital
ICU length of stay	Number of days spent in the ICU
First care unit	ICU type in which the patient was cared for
Admission type	Admission type the patient entered

Vital signs are clinical measurements that describe the state of a patient's body functions.

Vital sign	Description
Heart Rate	Heartbeat rate of the patient

Mean Blood Pressure	Average pressure in a patient's arteries during one cardiac cycle
Diastolic blood pressure	Pressure when the heart is at rest between beats
Systolic blood pressure	Pressure when the heart is beating
Respiratory Rate	Number of breaths taken per minute
Temperature	Temperature of a patient in degrees Celsius
SpO2	Amount of oxygen in the blood
Glasgow Coma Scale	Scoring system used to describe the level of consciousness in a person
Ventilation	Whether the patient was ventilated or not
Urine output	How much urine was produced

During a patients' stay in the ICU, their routine vital signs and additional information relevant to their care are constantly monitored and recorded electronically. This results in data taking the form of a time series where there is an ordered sequence of observations of many variables. I extracted every vital sign data point of the first 24 hours of a patient upon entering the ICU. The raw values of a few vital signs of a patient looks as follows:

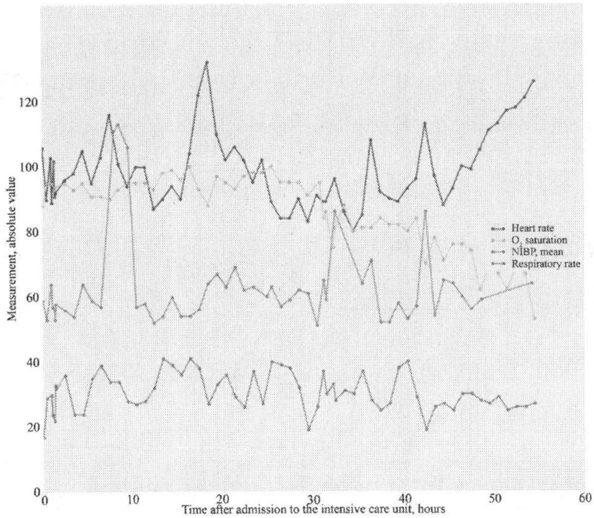

Laboratory measurements are made by acquiring a fluid from the patient's body (e.g., blood from an arterial line or urine from a catheter) and then analyzing it in the laboratory.

Laboratory measurements
Aniongap

Bicarbonate
Creatinine
Chloride
Glucose
Hematocrit
Hemoglobin
Platelet
Potassium
Sodium
Blood urea nitrogen
White blood cells

Laboratory measurements taken from a patient are also strong indicators of a patient's health condition. Let's take anion gap as an example. Anion gap is the difference between primary measured cations (sodium Na+ and potassium K+) and the primary measured anions (chloride Cl^- and bicarbonate HCO_3^-) in serum (blood). The test is mostly performed in patients with altered mental status, unknown exposures, acute renal failure, and acute illnesses. A kernel density estimation plot is used to view the distribution of the values below shows the aniongap measurement on ICU admission comparison for survival and non-survival groups.

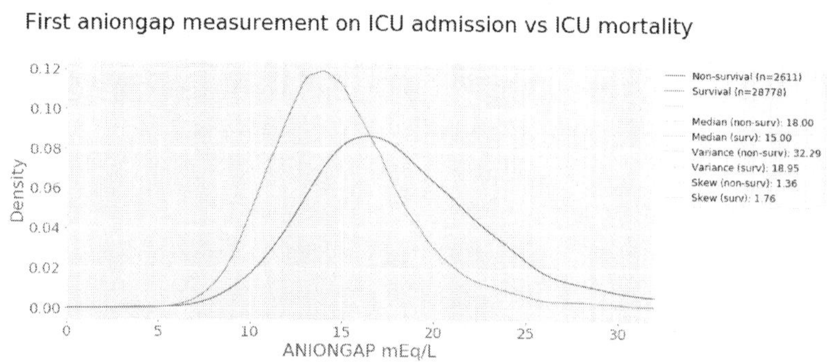

Let us get involved a little bit more into particulars.

```
In [1]:
# Import libraries
import numpy as np
import pandas as pd
import matplotlib.pyplot as plt
import psycopg2
import getpass
```

```
from modshogun import *
# below imports are used to print out pretty pandas
dataframes
from IPython.display import display, HTML
%matplotlib inline
plt.style.use('ggplot')
In [2]:
query_output = pd.read_csv('./full-features-mate-
rialized.csv') #.dropna().reset_index(drop=True)
query_output.first_careunit = pd.Categorical(query_
output.first_careunit)
query_output.gender = pd.Categorical(query_output.
gender)
query_output.admission_type = pd.Categorical(query_
output.admission_type)
query_output['gender'] = query_output.gender.cat.
codes
query_output['first_careunit'] = query_output.first_
careunit.cat.codes
query_output['admission_type'] = query_output.ad-
mission_type.cat.codes
query_output = query_output[(query_output.exclu-
sion_los == 0) & (query_output.exclusion_age == 0)
                              & (query_output.exclu-
sion_first_stay == 0) ]#& (query_output.age_icu_in
< 250)]
# Median age of patients > 89 is 91.6
query_output.loc[query_output.age_icu_in > 89,
'age_icu_in'] = 91.6
mortality_names = ['hospital_expire_flag', 'icu_ex-
pire_flag',
                  'hospital1year_expire_flag', 'hos-
pital30day_expire_flag']
los_names = ['hosp_los', 'icu_los']
feature_names = [
            # Demographic information
           'age_icu_in', 'gender', 'first_careunit',
            'admission_type',
```

```
            # Vital signs
            'heartrate_min', 'heartrate_max',
            'meanbp_min', 'meanbp_max',
            'diasbp_min', 'diasbp_max',
            'sysbp_min', 'sysbp_max',
            'resprate_min', 'resprate_max',
            'mingcs', 'gcsmotor',
            'gcsverbal', 'gcseyes',
            'tempc_min', 'tempc_max',
            'spo2_min', 'spo2_max',

            # Laboratory measurements

            'aniongap_min', 'aniongap_max',
            #'albumin_min', 'albumin_max',
            #'bands_min', 'bands_max',
            'bicarbonate_min', 'bicarbonate_max',
            #'bilirubin_min', 'bilirubin_max',
            'creatinine_min', 'creatinine_max',
            'chloride_min', 'chloride_max',
            'glucose_min', 'glucose_max',
            'hematocrit_min', 'hematocrit_max',
            'hemoglobin_min', 'hemoglobin_max',
            #'lactate_min', 'lactate_max',
            'platelet_min', 'platelet_max',
            'potassium_min', 'potassium_max',
            #'ptt_min', 'ptt_max',
            #'inr_min', 'inr_max',
            #'pt_min', 'pt_max',
            'sodium_min', 'sodium_max',
            'bun_min', 'bun_max',
            'wbc_min', 'wbc_max',

            # Other
            'vent',
            'urineoutput'
            ]
query_output = query_output.loc[:, mortality_names
+ feature_names]
```

```
query_output = query_output.fillna(query_output.
mean()).reset_index(drop=True)
# query_output = query_output.loc[:, mortality_names
+ feature_names].dropna().reset_index(drop=True)
outcomes = query_output.loc[:, mortality_names]
features = query_output.loc[:, query_output.col-
umns.difference(mortality_names + los_names)]
X = features
y = outcomes['hospital_expire_flag'].replace(0, -1)
print(len(query_output))
query_output.tail()
32290
Out[2]:
```

	hospital_expire_flag	icu_expire_flag	hospital1year_expire_flag	hospital30day_expire_flag	age_icu_in	gender	first_careunit	admission_type	heartrate_min	heartrate_max	...	potassium_min	potassium_max	sodium_min	sodium_max	bun_min	bun_max	wbc_min	wbc_max	vent	urineoutput
32285	0	0	0	0	70.826046	0	5	0	62.0	119.0	...	3.1	3.9	135.0	144.0	7.0	7.0	3.3	4.5	1	1995.0
32286	0	0	0	0	63.893494	0	2	1	60.0	95.0	...	3.5	3.9	139.0	143.0	16.0	31.0	5.3	8.6	0	1400.0
32287	0	0	0	0	43.594414	0	0	1	58.0	80.0	...	3.8	4.1	139.0	140.0	12.0	13.0	11.7	11.8	0	1250.0
32288	1	1	1	1	86.957864	1	2	1	52.0	68.0	...	5.0	5.7	123.0	128.0	41.0	42.0	11.4	13.3	1	730.0
32289	0	0	0	0	77.488270	1	0	1	28.0	74.0	...	4.4	4.9	135.0	138.0	21.0	22.0	10.8	11.0	0	1800.0

5 rows × 52 columns

6.5. PRINCIPAL COMPONENT ANALYSIS

PCA is used to reduce the dimensionality of your data. Without going into the theoretical aspects of PCA, we take a practical route by resorting to learning through writing appropriate code.

```
In [3]:
split = int(len(X) * 0.7)
X_train = RealFeatures(np.array(X[:split].T))
X_test = RealFeatures(np.array(X[split:].T))
y_train = BinaryLabels(np.array(y[:split]))
y_test = BinaryLabels(np.array(y[split:]))
print("Number of training samples:", y_train.get_
num_labels())
print("Number of testing samples:", y_test.get_num_
labels())
('Number of training samples:', 22603)
('Number of testing samples:', 9687)
In [4]:
preprocessor = PruneVarSubMean(True)
preprocessor.init(X_train)
X_train.add_preprocessor(preprocessor)
X_train.apply_preprocessor(True)
preprocessor.init(X_test)
X_test.add_preprocessor(preprocessor)
X_test.apply_preprocessor(True)
Out[4]:
True
In [5]:
preprocessor = PCA(AUTO)
preprocessor.set_target_dim(48)
preprocessor.init(X_train)
E = preprocessor.get_transformation_matrix()
mean = preprocessor.get_mean()
#Get all the eigenvalues returned by PCA.
eig_value = preprocessor.get_eigenvalues()
print(eig_value)
X_transformed = preprocessor.apply_to_feature_
matrix(X_train).T
print(X_transformed[:, 0])
```

```
[ 0.01518624   0.02105273   0.04639135   0.05442586
0.0594758   0.06029003
   0.09052768   0.12873543   0.13550631   0.21896065
0.22534858  0.22726647
   0.25028809   0.27740557   0.32681767   0.35340083
0.36384161  0.42265185
   0.4401412    0.44916387   0.47917281   0.51201246
0.57479038  0.63759838
   0.66896667   0.70007504   0.78275744   0.78933194
0.84620183  0.87661971
   0.90882884   0.95284362   0.99164835   1.01047059
1.18407776  1.26703612
   1.32142152   1.39900352   1.5517613    1.70586037
1.80320696  2.16376457
   2.41615497   2.60386819   2.71934855   3.6816201
4.34371495  4.94308891]
[ 0.01688562 -0.17268246 0.04270941 ..., 0.00933425
0.02306006
  0.0579221 ]
In [6]:
preprocessor = PCA(AUTO)
preprocessor.set_target_dim(2)
preprocessor.init(X_train)
E = preprocessor.get_transformation_matrix()
mean = preprocessor.get_mean()
#Get all the eigenvalues returned by PCA.
eig_value = preprocessor.get_eigenvalues()
print(eig_value)
#X_transformed   =   preprocessor.apply_to_feature_
matrix(X_train).T
#print(X_transformed[0, :])
[ 0.01518624   0.02105273   0.04639135   0.05442586
0.0594758   0.06029003
   0.09052768   0.12873543   0.13550631   0.21896065
0.22534858  0.22726647
   0.25028809   0.27740557   0.32681767   0.35340083
0.36384161  0.42265185
   0.4401412    0.44916387   0.47917281   0.51201246
0.57479038  0.63759838
```

```
  0.66896667  0.70007504  0.78275744  0.78933194
0.84620183  0.87661971
  0.90882884  0.95284362  0.99164835  1.01047059
1.18407776  1.26703612
  1.32142152  1.39900352  1.5517613   1.70586037
1.80320696  2.16376457
  2.41615497  2.60386819  2.71934855  3.6816201
4.34371495  4.94308891]
```

6.5.1. 2D Representation Using PCA

```
In [10]:
fig, plot = plt.subplots(figsize=(18, 12))
#fig.set_size_inches(25, 25)
plt.prism()
# X_transformed = preprocessor.apply_to_feature_
matrix(X_train).T # this crashes when ran twice
colors = [
    ('#2980b9', 'o'), # blue
    ('#c0392b', 'o'), # red
]
labels = np.unique(y_train.get_labels())
for num in range(len(labels)):
      plt.scatter([X_transformed[:,0][i] for i in
range(y_train.get_num_labels()) if y_train.get_
value(i) == labels[num]],
    [X_transformed[:,1][i] for i in range(y_train.
get_num_labels()) if y_train.get_value(i) ==
labels[num]], 25,
     label=['Survival', 'Non-Survival'][num], col-
or = colors[num][0], marker=colors[num][1])#,
edgecolors='black')

plt.legend(scatterpoints=1, fontsize=16)
plt.title("PCA Using the 1st and 2nd principal com-
ponent", fontsize=24)
#plot.set_xticks(())
#plot.set_yticks(())
plt.xlabel('PCA1', fontsize=16)
```

```
plt.ylabel('PCA2', fontsize=16)
plt.tight_layout()
```

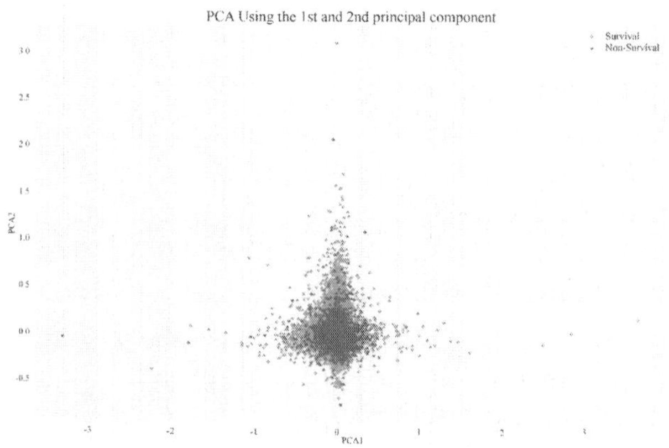

6.5.2. Retained Variance vs Performance

```
In [11]:
X_train = RealFeatures(np.array(X[:split].T))
X_test = RealFeatures(np.array(X[split:].T))
y_train = BinaryLabels(np.array(y[:split]))
y_test = BinaryLabels(np.array(y[split:]))
preprocessor = PruneVarSubMean(True)
preprocessor.init(X_train)
X_train.add_preprocessor(preprocessor)
X_train.apply_preprocessor(True)
preprocessor.init(X_test)
X_test.add_preprocessor(preprocessor)
X_test.apply_preprocessor(True)
preprocessor = PCA(AUTO)
preprocessor.set_target_dim(48)
preprocessor.init(X_train)
X_transformed = preprocessor.apply_to_feature_
matrix(X_train)
X_test_transformed = preprocessor.apply_to_fea-
ture_matrix(X_test)
n_components = [2, 4, 8, 10, 16, 20, 24, 32, 40,
44, 48]
rm = []
aucs = []
```

```
eigen_sum = sum(eig_value)
# Parameters to svm
C = 0.1
epsilon = 0.001
for component in n_components:

    X_train = X_transformed[:component, :]
    X_test = X_test_transformed[:component, :]
    X_train = RealFeatures(X_train)
    X_test = RealFeatures(X_test)

    lr = LibLinear(C, X_train, y_train)
    lr.set_liblinear_solver_type(L2R_LR)
    lr.set_epsilon(epsilon)
    lr.train()
    y_pred = lr.apply(X_test)
    lr_roc = ROCEvaluation()
    lr_roc.evaluate(y_pred, y_test)
    auc = lr_roc.get_auROC()
     print("Area under ROC(%):", auc, "(%d compo-
nents)" % component)

    aucs.append(auc)

    retained_variance = sum(eig_value[:component])
/ eigen_sum

    rm.append(retained_variance)
('Area under ROC(%):', 0.5126518085701792, '(2 com-
ponents)')
('Area under ROC(%):', 0.539211559619722, '(4 com-
ponents)')
('Area under ROC(%):', 0.5526297771195736, '(8 com-
ponents)')
('Area under ROC(%):', 0.5486884058312623, '(10
components)')
('Area under ROC(%):', 0.5664400664400573, '(16
components)')
('Area under ROC(%):', 0.5698240290077, '(20 com-
```

```
ponents)')
('Area under ROC(%):', 0.6002738553758937, '(24
components)')
('Area under ROC(%):', 0.6438200927996828, '(32
components)')
('Area under ROC(%):', 0.7111868540440038, '(40
components)')
('Area under ROC(%):', 0.764618172781434, '(44 com-
ponents)')
('Area under ROC(%):', 0.8467379691869136, '(48
components)')
In [159]:
plt.figure(figsize=(16, 12))
plt.plot(rm, aucs)
plt.xlabel("Retained variance")
plt.ylabel("auROC")
plt.title("AUC vs retained variance")
Out[159]:
<matplotlib.text.Text at 0x7f76a8757510>
```

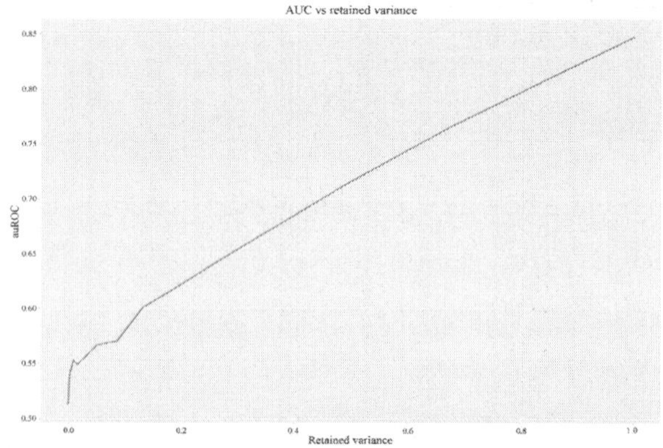

6.5.3. Isomap

```
In [87]:
%%time
# create instance of Isomap converter and configure
it
isomap = Isomap()
isomap.set_target_dim(2)
```

```
# set the number of neighbours used in kNN search
isomap.set_k(20)
# embed data
embedded_data_isomap = isomap.embed(X_train).get_
feature_matrix()
/usr/local/lib/python2.7/dist-packages/ipykernel_
launcher.py:9: RuntimeWarning: [WARN] In file /build/
shogun-v9ad6W/shogun-6.0.0+1SNAPSHOT201704270057/
src/shogun/lib/tapkee/tapkee_shogun.cpp line 35:
The neighborhood graph is not connected.
  if __name__ == '__main__':
CPU times: user 31min 52s, sys: 11.9 s, total: 32min
4s
Wall time: 15min 6s
In [90]:
fig, plot = plt.subplots(figsize=(18, 12))
#fig.set_size_inches(25, 25)
plt.prism()
# X_transformed = preprocessor.apply_to_feature_
matrix(X_train).T # this crashes when ran twice
colors = [
    ('#2980b9', 'o'), # blue
    ('#c0392b', 'o'), # red
]
labels = np.unique(y_train.get_labels())
for num in range(len(labels)):
   plt.scatter([embedded_data_isomap.T[:,0][i] for
i in range(y_train.get_num_labels()) if y_train.
get_value(i) == labels[num]],
   [embedded_data_isomap.T[:,1][i] for i in range(y_
train.get_num_labels()) if y_train.get_value(i) ==
labels[num]], 25,
     label=['Survival', 'Non-Survival'][num], col-
or = colors[num][0], marker=colors[num][1])#,
edgecolors='black')

plt.legend(scatterpoints=1, fontsize=16)
plt.title("Isomap 2D Plot", fontsize=24)
#plot.set_xticks(())
```

```
#plot.set_yticks(())
plt.tight_layout()
```

6.6. *T*- DISTRIBUTED STOCHASTIC NEIGHBOR EMBEDDING (T-SNE)

t-Distributed Stochastic Neighbor Embedding (t-SNE) is a technique for dimensionality reduction that is particularly well suited for the visualization of high-dimensional datasets. The technique can be implemented via Barnes-Hut approximations, allowing it to be applied on large real-world datasets. t-SNE is a tool to visualize high-dimensional data. It converts similarities between data points to joint probabilities and tries to minimize the Kullback-Leibler divergence between the joint probabilities of the low-dimensional embedding and the high-dimensional data. t-SNE has a cost function that is not convex, i.e. with different initializations we can get different results. It is highly recommended to use another dimensionality reduction method (e.g., PCA for dense data or Truncated SVD for sparse data) to reduce the number of dimensions to a reasonable amount (e.g., 50) if the number of features is very high. This will suppress some noise and speed up the computation of pairwise distances between samples.

It is a nonlinear dimensionality reduction technique that is particularly well-suited for embedding high-dimensional data into a space of two or three dimensions, which can then be visualized in a scatter plot. Specifically, it models each high-dimensional object by a two- or three-dimensional point in such a way that similar objects are modeled by nearby points and dissimilar objects are modeled by distant points.

The t-SNE algorithm comprises two main stages.

1. t-SNE constructs a probability distribution over pairs of high-dimensional objects in such a way that similar objects have a high probability of being picked, whilst dissimilar points have an extremely small probability of being picked.
2. t-SNE defines a similar probability distribution over the points in the low-dimensional map, and it minimizes the Kullback–Leibler divergence between the two distributions with respect to the locations of the points in the map. Note that whilst the original algorithm uses the Euclidean distance between objects as the base of its similarity metric, this should be changed as appropriate.

```
In [11]:
tsne = TDistributedStochasticNeighborEmbedding()
tsne.set_target_dim(2)
tsne.set_perplexity(25)
embedded_data_tsne = tsne.embed(X_train).get_fea-
ture_matrix()
In [12]:
fig, plot = plt.subplots(figsize=(18, 12))
#fig.set_size_inches(25, 25)
plt.prism()
# X_transformed = preprocessor.apply_to_feature_
matrix(X_train).T # this crashes when ran twice
colors = [
    ('#2980b9', 'o'), # blue
    ('#c0392b', 'o'), # red
]
labels = np.unique(y_train.get_labels())
for num in range(len(labels)):
     plt.scatter([embedded_data_tsne.T[:,0][i] for
i in range(y_train.get_num_labels()) if y_train.
get_value(i) == labels[num]],
    [embedded_data_tsne.T[:,1][i] for i in range(y_
train.get_num_labels()) if y_train.get_value(i) ==
labels[num]], 25,
     label=['Survival', 'Non-Survival'][num], col-
or = colors[num][0], marker=colors[num][1])#,
edgecolors='black')

plt.legend(scatterpoints=1, fontsize=16)
plt.title("t-SNE 2D Plot", fontsize=24)
#plot.set_xticks(())
#plot.set_yticks(())
plt.xlabel('Dimension 1', fontsize=16)
plt.ylabel('Dimension 2', fontsize=16)
plt.tight_layout()
```

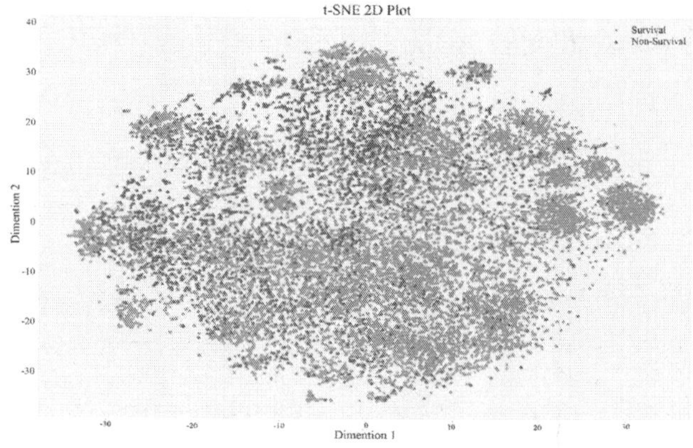

6.7. VISUALIZATION

To get a better grasp of the effects of the predictors on the mortality outcomes, I explored the dataset with multiple visualizations. Here, I used a 3d plot to see how different features affect a patient's mortality probability.

In this visualization, I included age, heart rate and the Glasgow Coma Scale which is a score indicating the level of consciousness of a person. From the plot, patients with somewhat extreme values (low heart rate and Glasgow Coma Scale values) are more likely to die, shown with a red 'X.' Although this plot only shows 3 predictors, it is possible to change the variables on the 3 axis for visualizations.

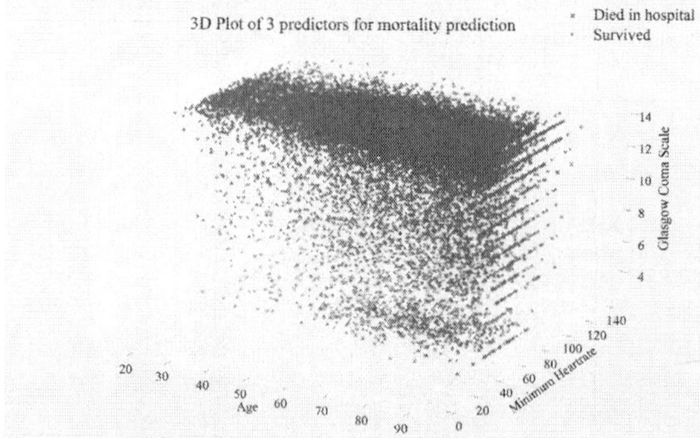

In total, 48 features are used to build the model for both prediction tasks. Since there are a high number of features, I was interested in seeing how high-dimensional data would look like in 2D using dimensionality reduction techniques like PCA and t-SNE.

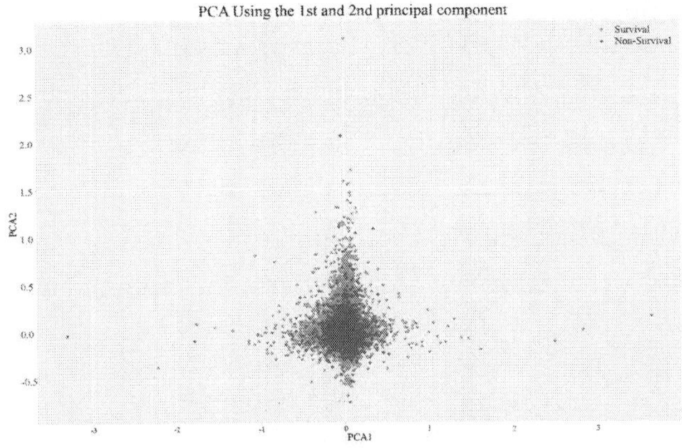

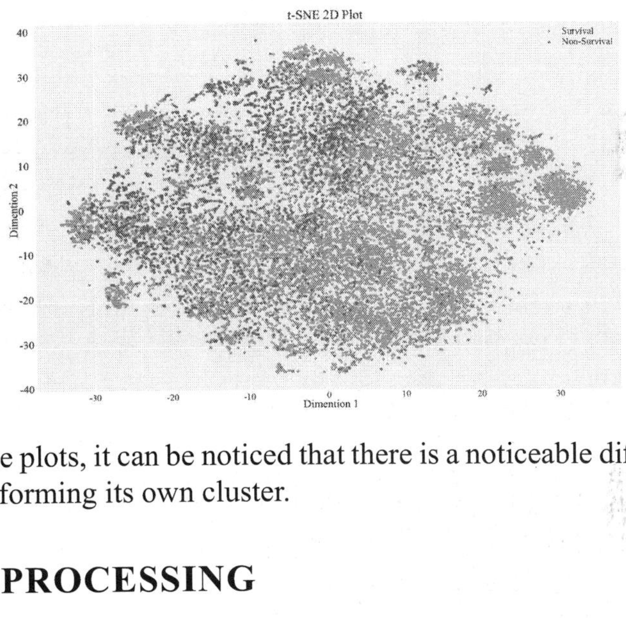

From the plots, it can be noticed that there is a noticeable difference with each group forming its own cluster.

6.8. PREPROCESSING

6.8.1. Exclusions

Because MIMIC is an ICU database, the focus was placed on patients admitted to and discharged from the ICU. Patients admitted to the ICU generally suffer from severe and life-threatening illnesses and injuries which require constant, close monitoring and support. Being able to make good decisions during this time period is therefore crucial. For that reason, data points were queried and grouped based off the ICU stay rather than the individual patient to develop a model specifically for ICU patient monitoring and decision-making.

The selection criteria is described below along with a short explanation. The following points were excluded from the dataset:

- Patients aged less than 16 years old

This also removed newborns and children, which likely have different predictors than adults

- Second admissions of patients
- Simplifies analysis which assumes independent observations

  We avoid taking into account that ICU stays are highly correlated
- Length of stay less than 1 day

  Helps remove false positives that were placed in ICU for precautionary purposes

6.8.2. Data Cleaning

Because not all lab measurements are recorded for every patient, I replaced missing values and NaNs in the dataset with the mean value as a data imputation technique, to maintain a reasonable size for the training and testing set and not to introduce any bias.

Additionally, data standardization was applied to make each feature have zero mean by subtracting the mean, and have unit-variance to ensure that all the data is normalized, that the features are in the same range. This also speeds up the training time for classifiers like the SVM.

6.8.3. Basic Model

In this first model, I investigated how well predictions could be when not using time series data. I ignored the temporal structure of the data, and only used the entire data of the first 24 hours in the ICU for a patient. A single time-shot computing the minimum, maximum and mean for that time period was used for the vital signs. For mortality prediction, three machine-learning classifiers were used: random forest, logistic regression and linear support vector machine. These algorithms are commonly used and allow to learn the relationship between predictor variables and a binary outcome variable.

6.9. RESULTS

One can evaluate the performance of a model using the auROC as the evaluation metric for mortality prediction and recorded the results in the table below using stratified 10-fold cross-validation. Stratified 10-fold cross-validation gives a good estimate for the performance of our model, as the data is divided into *k*- equally sized folds in such a way that labels in each partition are roughly the same. I also compare the scores with *sklearn's* implementation of logistic regression and linear SVM and yielded identical results. Finally, I compared the result with XGBoost, which is a popular algorithm used in Kaggle competitions.

Classifier	Mean AUC across 10 folds (%)
	Hospital mortality
Logistic Regression	84.64
Linear SVM	84.56
Logistic Regression (sklearn)	84.64
Linear SVM (sklearn)	84.56
XGBoost	87.60
Random guess	50.00

For all three classifiers, 1-day and 1-week mortality predictions were evaluated. The barcharts below show the results for the different tasks.

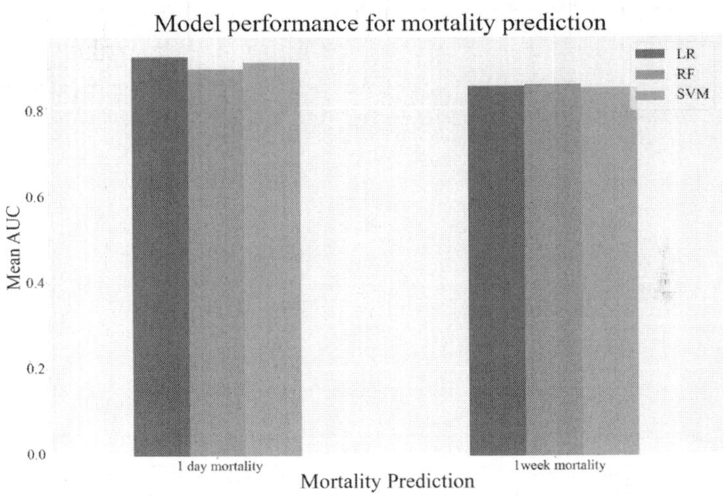

Boxplots give an indication of the variance of the results over the 10 folds through cross-validation.

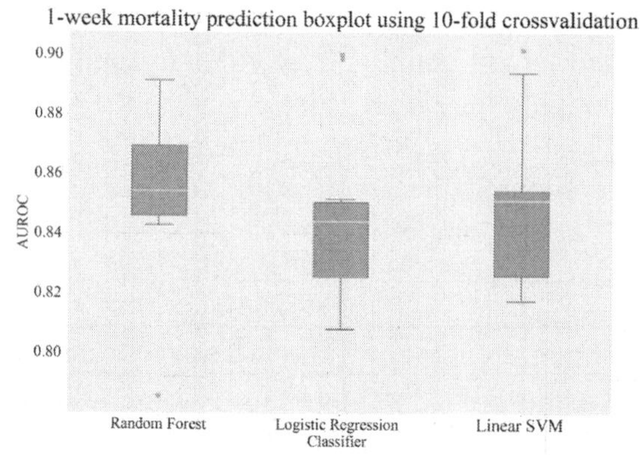

The ROC curve for hospital mortality gives an insight on the sensitivity and specificity of our logistic regression and linear SVM models. The performance of both models were quite similar.

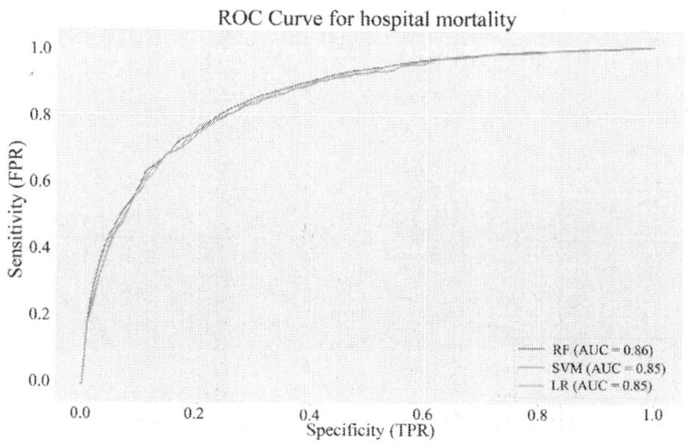

Finally, the regression task for predicting hospital length of stay was evaluate using mean squared error.

Classifier	*Mean MSE across 10 folds*	
	Hospital length of stay	ICU length of stay
Least square regression	110.726	46.91
Linear ridge regression	110.726	46.91

6.10. IMPROVED MODEL WITH TEMPORAL FEATURES

The previous basic model above that was built for mortality prediction was based on a single time-shot features i.e. vital signs of the first 24 hours were aggregated and had descriptive statistics computed. As a result, the basic model fails to account any temporal structure in the data and will not capture a changing health status of a patient.

Let's recall that the objective was to develop a model that could allow caregivers to monitor and predict mortality of patients. In practice, such a system would be incorporated in a real-time clinical monitoring system where it would be possible to look at a patient at any point in time and make a prediction about whether the patient will die within a certain amount of time. The previous model is limited in the sense that it doesn't account for predictions at any point in time, since it's trained with data from only the first 24 hours combined.

In this improved model, I exploit the temporal information, taking into account the fluctuations of the vital signs over time and most importantly does not look at how the vital signs change moments before death. From there, it allowed me to better understand how a dying patient differs from a patient with normal behavior moments before days, hours and right before death.

6.10.1. Resampling

One of the challenges with the MIMIC database is that vital signs are measured and recorded at irregular time intervals. Resampling refers to converting raw time series data into discrete intervals at a fixed frequency. A smaller time resolution between vital sign readings, the better and more accurate the classifier becomes since it allows for timely predictions.

Because space and computing power is limited, a sampling rate measuring capturing vital signs every hour was chosen, for the first 24 hours of a patient. Afterwards, lagged features from the last 3 hourly observations (t–1, t–2, t–3), which are the vital signs at the previous hours, were computed by shifting columns of the time stamp.

A rolling window then computes the mean, min, max and median over the entire vital signs time series of a patient in the last 6 hours. The same vital signs used previously were included for this model. Below is a sample of data points with only heart rate used.

time	hr	hr_1h	hr_max_6h	hr_mean_6h	hr_median_6h	hr_min_6h
13:00	90.0	74.0	90.0	79.85	77.0	74.0
14:00	81.0	90.0	90.0	79.28	77.0	74.0
15:00	79.0	81.0	90.0	79.00	77.0	74.0
16:00	79.0	79.0	90.0	79.42	79.0	74.0
17:00	68.0	79.0	90.0	78.14	79.0	68.0

To observe how these features differ between dying and surviving patients, I plotted stacked histograms from equally sampled survival and non-survival patients to view the distribution of these vital signs values across both groups.

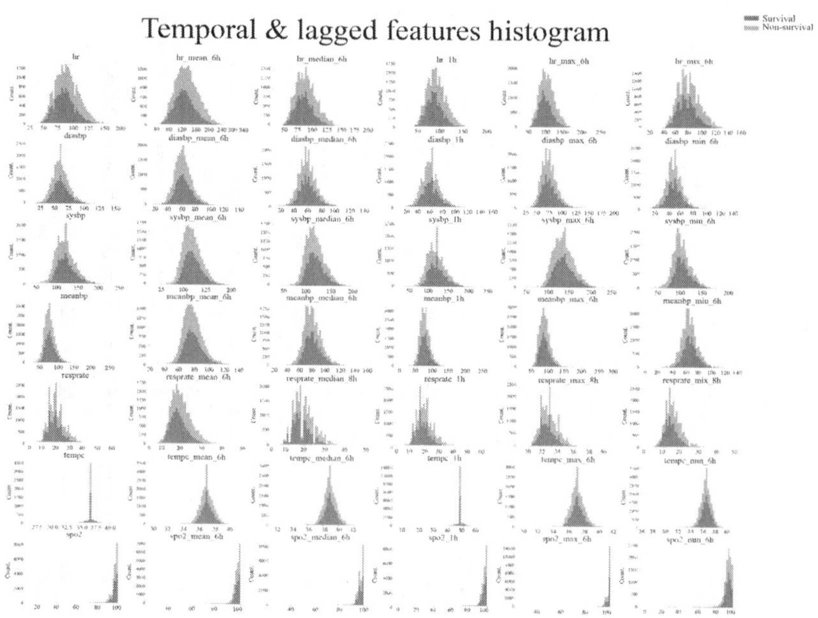

While there is no clear conclusion that can be drawn from looking at the vital signs histograms, this may indicate that these variables vary depending on the patient. For instance, the mean/median value of heart rate greatly varies depending on the patient, even when dying.

6.10.2. Model

In the end, the features included vital signs, their lagged observations and window computations, demographic information and the hospital length of stay of the patient. The feature matrix consisted of a single row represented a patient observation at a certain point in time with the features mentioned above. The following plot shows what sample features look like for a dying patient.

Temporal features plot for a dying patient

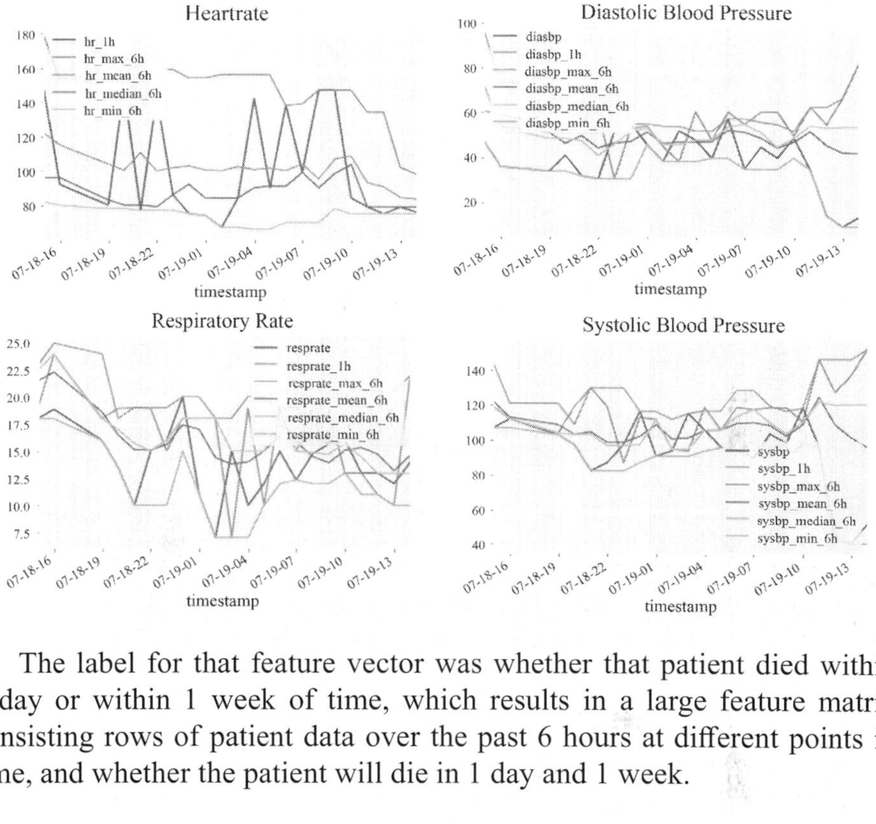

The label for that feature vector was whether that patient died within 1 day or within 1 week of time, which results in a large feature matrix consisting rows of patient data over the past 6 hours at different points in time, and whether the patient will die in 1 day and 1 week.

6.10.3. Results

Three classifiers (logistic regression, linear SVM and random forest) were evaluated and compared. One challenge I faced was dealing with class imbalance where the number of negative samples (patients that did not die) was significantly larger than the number of positive samples (patients that died). This was addressed by under sampling the negative samples to maintain a ratio of 80/20. In addition, because the size of the dataset consisted of over 500,000 data points, the training and testing set were subsampled and evaluated several times. The subset of the data consisted of 100,000 points and split in two for training and testing, and averaged over 10 runs. The results were recorded in the following table.

Classifier	*Test set AUC score (%)*	
	1-day hospital mortality	1-week hospital mortality
Logistic Regression	73.48	72.03
Linear SVM	73.22	71.81
Random Forest	83.12	83.51

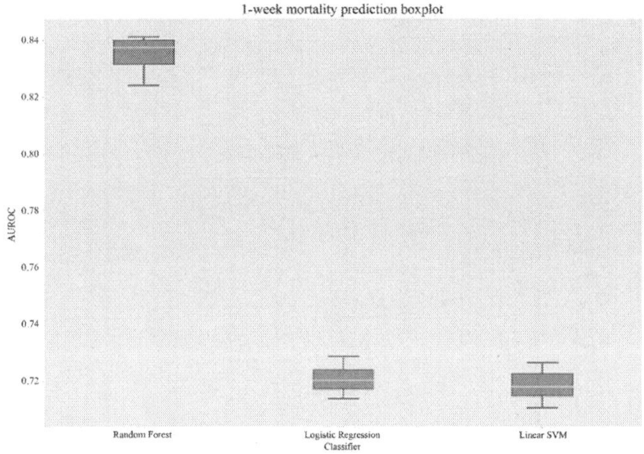

Random forest produced the best results among all three classifiers for 1-day and 1-week mortality prediction. The variance across the 10 runs with 100,000 samples from the entire dataset can also be seen here.

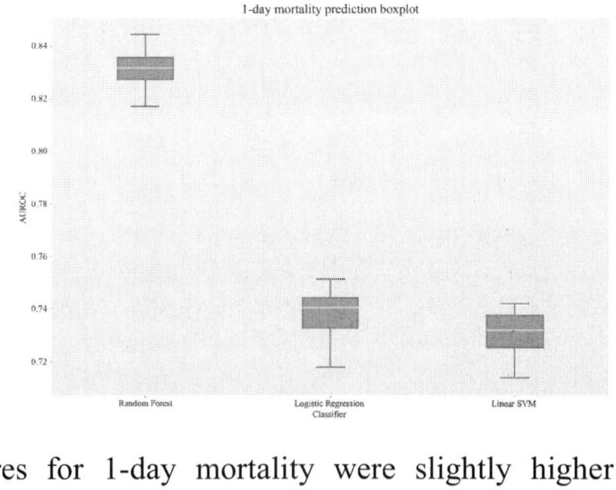

The scores for 1-day mortality were slightly higher than 1-week mortality, which could be explained due to the fact that the first 24 hours upon entering the ICU are critical and are closer in time to the prediction period of 1 day.

6.11. CONCLUSIONS

We started off with a basic model using demographic information, vital signs and laboratory measurements for predicting patient mortality. Using three common machine learning classifiers, one was able to obtain decent results of 85% AUC. However, ignoring the temporal component of the data reduced the predictive power of models, since a single time-shot of all vital signs to predict mortality was employed. In the second, improved model, we added lagged features and computed sliding window summary statistics for the vital signs. The latter model ended up being much more realistic and capable of making better predictions at any point in time since it took into account the time series data.

REFERENCES

1. Ross, M. K., Wei Wei, & L. Ohno-Machado. (2014). "Big data" and the electronic health record." Yearbook of Medical Informatics 9(1), 97.
2. Johnson, Alistair, E. W., et al. (2016). "MIMIC-III, a freely accessible critical care database." Scientific data 3

7 CHAPTER

BIOMEDICAL IMAGE PROCESSING

CONTENTS

Image Processing is an established field in Engineering sciences. In fact, Biomedical Image Processing (BIP) is also a ver well known area in biomedical sciences, with strong theoretical and computational foundation. It is a very broad field; and basically covers biomedical signal gathering, image forming, picture processing, and image display to medical diagnosis based on features extracted from images. Many of the concepts relevant to BIP are borrowed from the computational and engineering field of Image Processing. We introduce briefly the fundamentals of BIP in this chapter and switch to applications.

7.1. INTRODUCTION

The field of biomedicine relies heavily on direct digital imaging systems for medical diagnostics. Digital Image Processing (DIP) has gained prime importance in healthcare during the past three to four decades. Computed Tomography (CT) and Magnetic Resonance Imaging (MRI) are considered as the primary digital methods. Endoscopy and Radiography now come with sensors which are digital in nature. The digital images are composed of pixels (an acronym for "picture" and "element"). One assigns brightness or color variation to each such pixel. Let us now see what are the core steps in a DIP.

7.1.1. Steps in DIP

Instead of restricting over selves to BIP, we will see in this section the primary steps involved in any digital image processing. We mention them below:

a) Image formation: Capture an image. Create a digital matrix
b) Image Visualization: Manipulate the previous digital matrix to generate an optimized image output
c) Image Analysis: This is the core stp involving processing, quantification and abstract interpretation of images.
d) Image Management: this involves storage, communication, transmission of data.

All these core concepts are discussed in great detail in standard textbooks of Image Processing. Interested reader is referred to any one of the books which detail the above steps. We are going to take a practical approach towards computational techniques which can be applied to biomedical image data. So, let us assume that we have image data and proceed with computation.

7.2. PYTHON LIBRARY: BIOVIDA

We introduce a Python library called BioVida, which is designed to make the biomedical image data accessible to computational scientists for application of machine learning (ML) methods to gain insights into human disease.

7.2.1. How to Obtain Biovida?

Install it using pip as follows:

pip install biovida

You will be surprised that by using just a few lines of code, one can gain access to biomedical databases which store tens of millions of images. Please make sure that you adhere to to the copyright and usage restrictions of the respective repositories.

BioVida requires: beautiful soup, h5py, keras, lxml, numpy, pandas, pillow, pydicom, requests, scikit-image, scipy, theano and tqdm.

The installer will automatically install all of these packages.

A few repositories are described in the sections below.

7.3. OPEN-I

The Open-i^{sm} (pronounced "open eye") experimental multimedia search engine retrieves and displays structured MEDLINE citations augmented by image-related text and concepts and linked to images based on image features.

The Open-i project provides novel information retrieval services for biomedical articles from the full text collections such as PubMed Central. It is unique in its ability to index both the text and images in the articles. The article retrieval is powered by the LHNCBC-built search engine Essie.

The Open-I biomedical image search engine lets users retrieve not only the MEDLINE citation information, but also the outcome statements in the article and the most relevant figure from it. Further, it is possible to use the figure as a query component to find other relevant images or other visually similar images. Future stages aim to provide image region-of-interest (ROI) based querying. The initial number of images is projected to be around 600,000 and will scale to millions. The extensive image analysis and indexing and deep text analysis and indexing require distributed computing. Future plans include making the image computation services available as a NLM service.

7.4. THE CANCER IMAGING ARCHIVE (TCIA)

The image data in The Cancer Imaging Archive (TCIA) is organized into purpose-built collections of subjects. The subjects typically have a cancer type and/or anatomical site (lung, brain, etc.) in common. Each link in the table below contains information concerning the scientific value of a collection, information about how to obtain any supporting non-image data which may be available, and links to view or download the imaging data. To support reproducibility in scientific research, TCIA supports Digital Object Identifiers (DOIs) which allow users to share subsets of TCIA data referenced in a research manuscript. As of today, the data which is available from TCIA can be summarized in the table below (Table 7.1). So, let us get started with a bit of coding in BioVida to access the data.

Open-i BioMedical Image Search Engine

```
# 1. Import the Interface for the NIH's Open-i API.
from biovida.images import OpeniInterface
# 2. Create an Instance of the Tool
opi_1 = OpeniInterface()
# 3. Perform a search for x-rays of breast cancer
opi_1.search(query='breast cancer', image_type=['x_
ray'])  # Results Found: 2,025.
# 4. Pull the data
search_df = opi_1.pull()
```

See how simple it is to load the required data.Now, let us do the same TCIA.

Cancer Imaging Archive

```
# 1. Import the interface for the Cancer Imaging Archive
from biovida.images import CancerImageInterface
# 2. Create an Instance of the Tool
cii_1 = CancerImageInterface(YOUR_API_KEY_HERE)
# 3. Perform a search
cii_1.search(cancer_type='esophageal')
```

# 4. Pull the data
cdf = cii.pull()

Collection	Cancer Type	Modalities	Subjects	Location	Metadata	Access	Status	Updated
Brain-Tumor-Progression	Brain Cancer	MR	20	Brain	No	Limited	Complete	2018/01/31
Acrin-FLT-Breast	Breast Cancer	PET, CT, OT	83	Breast	Yes	Limited	Complete	2017/12/11
NSCLC-Radiogenomics	Non-small Cell Lung Cancer	PT, CT, SEG, SR	211	Chest	Yes	Public	Complete	2017/12/04
Head-Neck-PET-CT	Head and Neck Cancer	PT, CT, RTSTRUCT, RTPLAN, RTDOSE	298	Head-Neck	Yes	Public	Complete	2017/11/27
HNSCC	Head and Neck Squamous Cell Carcinoma	CT, PT, MR, RT-STRUCT, RTPLAN, RTDOSE	215	Head-Neck	Yes	Public	Complete	2017/10/31
TCGA-BLCA	Bladder Endothelial Carcinoma	CT, CR, MR, PT	106	Bladder	Yes	Public	Ongoing	2017/10/30
LGG-1p19qDeletion	Low Grade Glioma	MR	159	Brain	Yes	Public	Complete	2017/09/30
LCTSC	Lung Cancer	CT, RTSTRUCT	60	Lung	Yes	Public	Ongoing	2017/09/28
CBIS-DDSM	Breast Cancer	MG	6671	Breast	Yes	Public	Complete	2017/09/27
Phantom FDA	Lung Phantom	CT	7	Phantom	No	Public	Complete	2017/08/30
QIN LUNG CT	Non-small Cell Lung Cancer	CT	47	Lung	Yes	Public	Complete	2017/07/31
CC-Radiomics-Phantom	Lung Phantom	CT, RTSTRUCT	17	Phantom	Yes	Public	Complete	2017/07/29
TCGA-HNSC	Head and Neck Squamous Cell Carcinoma	CT, MR, PT, RT-STRUCT, RTPLAN, RTDOSE	192	Head-Neck	Yes	Public	Ongoing	2017/03/30
PROSTATEx	Prostate Cancer	MR	346	Prostate	Yes	Public	Complete	2017/03/30
Mouse-Astrocytoma	Glioblastoma Multiforme	MR	48	Head	Yes	Public	Complete	2017/03/30
TCGA-LIHC	Liver Hepatocellular Carcinoma	MR, CT, PT	97	Liver	Yes	Public	Complete	2017/01/30
TCGA-LUSC	Lung Squamous Cell Carcinoma	CT, NM, PT	37	Lung	Yes	Public	Ongoing	2017/01/30
TCGA-KIRP	Kidney Renal Papillary Cell Carcinoma	CT, MR, PT	33	Renal	Yes	Public	Ongoing	2017/01/30
TCGA-LUAD	Lung Adenocarcinoma	CT, PT, NM	69	Chest	Yes	Public	Ongoing	2017/01/30
IvyGAP	Glioblastoma	MR, CT	39	Brain	Yes	Public	Complete	2016/12/30
Prostate Fused-MRI-Pathology	Prostate Cancer	MR	28	Prostate	Yes	Public	Complete	2016/11/30
ISPY1	Breast Cancer	MR, SEG	222	Breast	Yes	Public	Complete	2016/08/31
TCGA-PRAD	Prostate Cancer	CT, PT, MR	14	Prostate	Yes	Public	Complete	2016/08/29
NSCLC-Radiomics	Lung Cancer	CT, RTSTRUCT	422	Lung	Yes	Public	Ongoing	2016/05/20
MyelomaTT3PET	Myeloma	PET	300	Whole Body	Yes	Public	Coming Soon	2016/04/16

Pancreas-CT	Healthy Controls	CT	82	Pancreas	Yes	Public	Complete	2016/03/22
Breast-MRI-NACT-Pilot	Breast Cancer	MR, SEG	64	Breast	Yes	Public	Complete	2016/01/26
4D-Lung	Non-small Cell Lung Cancer	CT, RTSTRUCT	20	Lung	No	Public	Complete	2016/01/07
NRG-1308	Non-small Cell Lung Cancer	CT, RTSTRUCT, RTPLAN, RTDOSE	12	Lung	No	Limited	Complete	2015/09/14
RIDER Neuro MRI	Brain Cancer	MR	19	Brain	No	Public	Complete	2015/09/07
QIN-Breast	Breast Cancer	MR, PT, CT	67	Breast	Yes	Limited	Ongoing	2015/09/04
QIN-BRAIN-DSC-MRI	Low & High Grade Glioma	MR	49	Brain	No	Limited	Complete	2015/08/28
QIN GBM Treatment Response	Glioblastoma Multiforme	MR	54	Brain	No	Limited	Complete	2015/08/12
Soft-tissue-Sarcoma	Soft-tissue Sarcoma	FDG-PET/CT, MR, RTSTRUCT	51	Extremities	Yes	Public	Complete	2015/06/01
TCGA-UCEC	Uterine Corpus Endometrial Carcinoma	CT, CR, MR, PT	58	Uterus	Yes	Public	Ongoing	2015/04/29
Mouse-Mammary	Breast Cancer	MR	32	Abdomen	No	Public	Complete	2015/03/30
CT Lymph Nodes	Lymphade-nopathy	CT	176	Abdomen, Mediasti-num	Yes	Public	Complete	2015/03/16
TCGA-THCA	Thyroid Cancer	CT, PT	6	Thyroid	Yes	Public	Ongoing	2014/12/31
TCGA-BRCA	Breast Cancer	MR, MG	139	Breast	Yes	Public	Complete	2014/12/30
LungCT-Diagnosis	Lung Cancer	CT	61	Lung	Yes	Public	Complete	2014/12/30
TCGA-SARC	Sarcomas	CT, MR	5	Chest-Abdomen-Pelvis, Leg, TSpine	Yes	Public	Complete	2014/12/30
TCGA-CESC	Cervical Squamous Cell Carcinoma and Endocervical Adenocarci-noma	MR	54	Cervix	Yes	Public	Complete	2014/12/09
TCGA-ESCA	Esophageal Carcinoma	CT	16	Esophagus	Yes	Public	Complete	2014/12/09
TCGA-STAD	Stomach Ad-enocarcinoma	CT	46	Stomach	Yes	Public	Complete	2014/12/09
TCGA-OV	Ovarian Serous Cystadenocarci-noma	CT, MR	143	Ovary	Yes	Public	Complete	2014/11/05
TCGA-COAD	Colon Adeno-carcinoma	CT	25	Colon	Yes	Public	Complete	2014/10/14
TCGA-KIRC	Kidney Renal Clear Cell Carcinoma	CT, MR, CR	267	Renal	Yes	Public	Complete	2014/10/09
QIN-SARCOMA	Sarcomas	MR	15	Breast, Calf, Chest, Elbow, Knee, Leg, Shoulder, Thigh	No	Limited	Ongoing	2014/09/05

TCGA-LGG	Low Grade Glioma	MR, CT	199	Brain	Yes	Public	Complete	2014/09/04
QIN PET Phantom	PET Phantom	PT	2	Phantom	No	Public	Complete	2014/09/04
QIN-HeadNeck	Head and Neck Carcinomas	PT, CT, SR, SEG, RWV	156	Head-Neck	Yes	Public	Complete	2014/08/26
QIN Breast DCE-MRI	Breast Cancer	MR, KO	10	Breast	Yes	Public	Ongoing	2014/07/31
NSCLC-Radiomics-Genomics	Lung Cancer	CT	89	Lung	Yes	Public	Complete	2014/07/02
QIN Prostate	Prostate Cancer	MR	22	Prostate	No	Limited	Complete	2014/07/02
Lung Phantom	Lung Phantom	CT	1	Phantom	No	Public	Complete	2014/06/19
TCGA-KICH	Kidney Chromophobe	CT, MR	15	Kidney	Yes	Public	Complete	2014/05/21
TCGA-READ	Rectum Adenocarcinoma	CT, MR	3	Rectum	Yes	Public	Complete	2014/05/21
TCGA-GBM	Glioblastoma Multiforme	MR, CT, DX	262	Brain	Yes	Public	Complete	2014/05/08
SPIE-AAPM Lung CT Challenge	Lung Cancer	CT	70	Lung	Yes	Public	Complete	2014/01/02
Prostate-3T	Prostate Cancer	MR	64	Prostate	Yes	Public	Complete	2013/05/31
NaF Prostate	Prostate Cancer	PT, CT	9	Prostate	No	Public	Complete	2013/04/23
Head-Neck Cetuximab	Head and Neck Carcinomas	CT, PT, RTSTRUCT, RTPLAN, RTDOSE	111	Head-Neck	No	Public	Complete	2013/04/17
National Lung Screening Trial	Lung Cancer	CT	26254	Chest	Yes	Limited	Complete	2013/03/01
Prostate-Diagnosis	Prostate Cancer	MR	92	Prostate	Yes	Public	Complete	2013/01/30
LIDC-IDRI	Lung Cancer	CT, CR, DX	1010	Chest	Yes	Public	Complete	2012/03/21
QIBA CT-1C	Lung Phantom	CT, PR, SEG, SR	1	Phantom	No	Public	Complete	2011/11/28
RIDER Lung CT	Lung Cancer	CT	32	Chest	Yes	Public	Complete	2011/11/25
RIDER Phantom MRI	MRI Phantom	MR	10	Phantom	Yes	Public	Complete	2011/11/25
RIDER Phantom PET-CT	PET-CT Phantom	PT, CT	20	Phantom	Yes	Public	Complete	2011/11/25
RIDER Lung PET-CT	Lung Cancer	PT, CT	244	Lung	No	Public	Complete	2011/11/25
Breast Diagnosis	Breast Cancer	MR, PT, CT, MG	88	Breast	Yes	Public	Complete	2011/11/09
RIDER Breast MRI	Breast Cancer	MR	5	Breast	No	Public	Complete	2011/11/08
REMBRANDT	Low & High Grade Glioma	MR	130	Brain	Yes	Public	Complete	2011/11/03
CT Colonography	Colon Cancer	CT	825	Colon	Yes	Public	Complete	2011/10/31
Prostate-MRI	Prostate Cancer	MR	26	Prostate	Yes	Limited	Complete	2011/06/30

7.5. PROCESSING DATA FROM OPEN-I

Open-i is an open access biomedical search engine provided by the US National Institutes of Health. The service grants programmatic access to its over a million images through a RESTful web API. BioVida provides an easy-to-use python interface for this web API, located in the images subpackage.

```
In [1]:
from biovida.images import OpeniInterface
Using Theano backend.
In [2]:
opi = OpeniInterface()
```

As detailed in the previous section, one starts by creating an instance of the class. All BioVida interfaces accept at least two parameters: verbose and cache_path. The first simply determines whether or not the class provides you with additional updates as the class works. The second refers to the location where will be stored (or *cached*) on your computer. If left to its default, data will be cached in a directory entitled *biovida_cache* in your home directory. For most use cases, this should suffice.

7.5.1. Searching

To search the Open-i database, we can use the OpeniInterface's search method. To explore valid values that can be passed to search, we can use the options().

```
In [3]:
opi.options()
  - 'article_type'
  - 'collection'
  - 'exclusions'
  - 'fields'
  - 'image_type'
  - 'rankby'
  - 'specialties'
  - 'subset'
  - 'video'
```

The code above enumerates all of the parameters, apart from a specific query string, that can be passed to search(). Additionally, options() can be used to investigate the valid values for any one of these parameters as follows:

```
In [4]:
opi.options('collection')
  - 'history_of_medicine'
  - 'indiana_u_xray'
  - 'medpix'
```

```
  - 'pubmed'
  - 'usc_anatomy'
In [5]:
opi.options('image_type')
  - 'ct'
  - 'graphic'
  - 'microscopy'
  - 'mri'
  - 'pet'
  - 'photograph'
  - 'ultrasound'
  - 'x_ray'
```

Let's go ahead and perform a search for X-ray and CT images of breast cancer' from the PubMed collection/database.

```
In [6]:
opi.search(query='lung   cancer',   image_type=('x_
ray', 'ct'), collection='pubmed')
Results Found: 8,531.
```

7.5.2. Downloading Data

Now that we've defined a search, we can easily download some, or all, of the results found. For the sake of expediency, let's limit the number of results we download to the first 1000.

```
In [7]:
pull_df = opi.pull(download_limit=1000)
Number of Records to Download: 1,500 (chunk size:
30 records).
```

The text information associated with images are referred to as 'records', which are downloaded in 'chunks' of no more than 30 at a time.

Images, unlike records, are downloaded 'one by one.' However, pull() will check the cache before downloading an image, in an effort to reduce redundant downloads.

The dataframe generated by pull() can be viewed using either opi. records_db, or the pull_df used above to capture the output of pull(). Both will be identical. We can also view an abbreviated dataframe, opi.records_ db_short, which has several (typically unneeded) columns removed.

```
In [8]:
import numpy as np
def simplify_df(df):
    """This function simplifies dataframes
    for the purposes of this tutorial."""
    data_frame = df.copy()
     data_frame['cached_images_path'] = '/path/to/
image'
    return data_frame[0:5].replace({np.NaN: ''})
In [9]:
simplify_df(opi.records_db_short)
```

	mesh_major	mesh_minor	problems	abstract	affiliate	article_type	authors	cc_license	doc_source	image_caption	...	image_problems_from_text	imaging_modality_from_text	parsed_abstract	sex	modality_full	image_id_short	query	pull_time	cached_images_path	download_success
0			lung cancer	Lung cancer is one of the leading causes of ca...	Department of Nuclear Medicine and Molecular I...	other	Purandare NC, Rangarajan V	byncsa	PMC	(A-D) Nodal disease. Right upper paratracheal ...	...	(arrows, grids)	Computed Tomography (CT): chest		male	Computed Tomography (CT)	7	{'subset': None, 'rankby': None, 'collection':...	2017-04-10 15:24:37.098065	/path/to/image	True
1			nsclc	Non-small cell lung cancer (NSCLC) accounts fo...	Cardiopulmonary Department, Sant'Andrea Hospit...	research_article	Pezzuto A, Piraino A, Mariotta S	by	PMC	Case 2: (A) CT scan showing the lung after sur...	...		Computed Tomography (CT): chest		female	Computed Tomography (CT)	2	{'subset': None, 'rankby': None, 'collection':...	2017-04-10 15:24:37.098065	/path/to/image	True
2	(Adrenal Insufficiency/diagnosis*/drug therapy...	(Acute Disease, Adrenal Cortex Hormones/therap...	hypoxia; tachycardia	Background: Adrenal crisis after surgical pro...	Department of Orthopedic Surgery, Osaka Medica...	research_article	Naka N, Takenaka S, Nanno K, Moriguchi Y, Chun...	by	PMC	CT scan showing bilateral adrenal enlargement ...	...	(asterisks,)	Computed Tomography (CT): chest	{'background': 'Adrenal crisis after surgical ...	male	Computed Tomography (CT)	1	{'subset': None, 'rankby': None, 'collection':...	2017-04-10 15:24:37.098065	/path/to/image	True
3			small cell carcinoma of lung	Subcutaneous swelling as first clinical presen...	Department of Medicine.	case_report	Kumar S, Gupta A, Diwan SK, Bhake A		PMC	Computerized tomography of the chest showing p...	...		Computed Tomography (CT): chest		male	Computed Tomography (CT)	3	{'subset': None, 'rankby': None, 'collection':...	2017-04-10 15:24:37.098065	/path/to/image	True
4	(Echocardiography/methods*, Lung/ultrasonograp...	(Aged, Humans, Incidental Findings, Male)	lung tumor	We present images of a rare case where a prima...	Department of Clinical Physiology and Nuclear ...	research_article	Dencker M, Cronberg C, Damm S, Valind	by	PMC	Display of CT-image. The arrow indicates the n...	...	(arrows,)	Computed Tomography (CT): chest		male	Computed Tomography (CT)	2	{'subset': None, 'rankby': None, 'collection':...	2017-04-10 15:24:37.098065	/path/to/image	True

```
5 rows × 38 columns
```

This dataframe provides a lot of rich data, which is extremely valuable. This is independent of the images which have also been downloaded.

For instance, it is possible to quickly generate some descriptive statistics about our newly created 'lung cancer' dataset.

```
In [10]:
pull_df['age'].describe()
Out[10]:
count    966.000000
mean      53.337267
std       21.999990
min        1.000000
25%       47.000000
50%       59.000000
75%       68.000000
max       91.400000
Name: age, dtype: float64
In [11]:
pull_df['sex'].value_counts(normalize=True)
Out[11]:
male      0.722359
female    0.277641
Name: sex, dtype: float64
```

The age and sex columns are generated by analyzing the raw text provided by Open-i.

It should also be mentioned that opi.records_db *only* contains data for the most recent search() and pull(). Hence, keep a track of the session. Conversely, **cache_records_db** provides a more complete account of all images in the cache, e.g., those obtained several sessions ago. Additionally, unlike opi.records_db, cache_records_db can contain duplicate rows. However, this is only allowed to occur if the queries that generated the rows are different. Please make sure that you don't have duplicate records in your dataframe.

7.5.3. Images

Now that we've explored obtaining and reviewing data, we can finally turn our attention to images themselves.

```
In [12]:
from utils import show_image
%matplotlib inline
```

Note: utils is a small script with some helpful functions located in the base of this directory.

Using the show_images imported above, we can now look at a random images we pulled in the step above.

```
In [13]:
# show_image(opi.records_db['cached_images_path'].
iloc[156])
In [14]:
opi.records_db['license_type'].iloc[156]
Out[14]:
'open-access'
```

Let's also look at the age and sex of this subject.

```
In [15]:
age_sex = opi.records_db['age'].iloc[156], opi.re-
cords_db['sex'].iloc[156]
print("age: {0}, sex: {1}.."format(*age_sex))
age: 48.0, sex: male.
```

We can also easily check their diagnosis

```
In [16]:
opi.records_db['diagnosis'].iloc[156]
Out[16]:
'carcinoma; neurofibromatosis'
```

Please be advised that for collections other than 'MedPix'*, such as PubMed, diagnosis information is obtained by analyzing the text associated with the image. ***Errors are possible***.

*MedPix explicitly provides diagnosis information, so it can be assumed to be accurate.

7.5.4. Automated Cleaning of Image Data

While the data may look OK so far, if we look more closely we will likely find several problems with the images we have downloaded.

```
In [17]:
# show_image(opi.records_db['cached_images_path'].
iloc[100])
In [18]:
# show_image(opi.records_db['cached_images_path'].
iloc[10])
```

The images obtained above contain several clear problems. They both contain arrows and the latter is actually a 'grid' of images. These are liable to confuse any model we attempt to train detect disease. We could manually go through and remove these images or, alternatively, we can use the *experimental* OpeniImageProcessing class to try and eliminate these images from our dataset automatically.

```
In [19]:
from biovida.images import OpeniImageProcessing
```

We initialize this class using our OpeniInterface instance. By default, it will extract the records_db DataFrame. Do note, however, that we can force it to extract the cache_records_db DataFrame by setting the db_to_extract equal to 'cache_records_db.'

```
In [20]:
ip = OpeniImageProcessing(opi)
```

OpeniImageProcessing will automatically download a model for a *Convolutional Neural Network* (convnet) which has been trained to detect these kinds of problems. We have described CNNs in the context of Deep Learning in the previous chapter.

The OpeniImageProcessing class tries to detect problems in the images by analyzing both the text associated it is associated with as well as by feeding the image through the convnet mentioned above. However, by default the OpeniImageProcessing class will only use predictions gleaned from this model if it has been explicitly trained on images from that kind of imaging modality.

We can easily check the modalities for which the model has been trained:

```
In [21]:
ip.trained_open_i_modality_types
Out[21]:
['ct', 'mri', 'x_ray']
```

Luckily, we're working with X-rays and CTs.

Now we're ready to analyze our images.

```
In [22]:
analysis_df = ip.auto()
In [23]:
simplify_df(analysis_df)
Out[23]:
```

	mesh_major	mesh_minor	problems	abstract	affiliate	article_type	authors	cc_license	detailed_query_url	doc_source	...	grayscale	medpix_logo_bounding_box	hbar	hborder	vborder	upper_crop	lower_crop	visual_image_problems	invalid_image	invalid_image_reasons
0			lung cancer	Lung cancer is one of the leading causes of ca...	Department of Nuclear Medicine and Molecular I...	other	Purandare NC, Rangarajan V	byncsa	https://openi.nlm.nih.gov/retrieve.php?img=PMC...	PMC	...	False							[(grids, 0.864566), (arrows, 0.00571598), (tex...	True	(grayscale, image_problems_from_text, visual_i...
1			nsclc	Non-small cell lung cancer (NSCLC) accounts fo...	Cardiopulmonary Department, Sant'Andrea Hospit...	research_article	Pezzuto A, Piraino A, Mariotta S	by	https://openi.nlm.nih.gov/retrieve.php?img=PMC...	PMC	...	True							[(valid_image, 0.858156), (text, 0.100862), (a...	False	
2	(Adrenal Insufficiency/diagnosis*/drug therapy...	(Acute Disease, Adrenal Cortex Hormones/therap...	hypoxia; tachycardia	Background: Adrenal crisis after surgical pro...	Department of Orthopedic Surgery, Osaka Medica...	research_article	Naka N, Takenaka S, Nanno K, Moriguchi Y, Chun...	by	https://openi.nlm.nih.gov/retrieve.php?img=PMC...	PMC	...	True		327				327	[(valid_image, 0.914723), (arrows, 0.0369132),...	True	(image_problems_from_text,)
3			small cell carcinoma of lung	Subcutaneous swelling as first clinical presen...	Department of Medicine.	case_report	Kumar S, Gupta A, Diwan SK, Bhake A		https://openi.nlm.nih.gov/retrieve.php?img=PMC...	PMC	...	False		348	(4, 351)	(191, 491)	4	348	[(arrows, 0.998476), (grids, 0.000149499), (va...	True	(grayscale, visual_image_problems)
4	(Echocardiography/methods*, Lung/ultrasonograp...	(Aged, Humans, Incidental Findings, Male)	lung tumor	We present images of a rare case where a prima...	Department of Clinical Physiology and Nuclear ...	research_article	Dencker M, Cronberg C, Damm S, Valind S, Wadbo M	by	https://openi.nlm.nih.gov/retrieve.php?img=PMC...	PMC	...	False		294		(96, 487)		294	[(grids, 0.993127), (arrows, 8.44916e-06), (va...	True	(grayscale, image_problems_from_text, visual_i...

5 rows × 65 columns

This will generate several new columns, which are described below:

- **'grayscale'**: this is simply an account of whether or not the images is grayscale.
- **'medpix_logo_bounding_box'**: images from the MedPix collection, typically contain the organization's logo in the top right corner. Had we passed the class images from MedPix, it would have tried to 'draw' a bounding box around its precise location (enabling it to be cropped out of the image).
- **'hbar'**: this denotes a 'horizontal bar' that is sometimes found at the bottom of images. If present, this column reports its height in pixels.
- **'hborder'**: this column provides an account of 'horizontal borders' on either side of the image.
- **'vborder'**: this column provides an account of 'vertical borders' on the top and bottom of the image.
- **'upper crop'**: this is the location that has been selected to crop the top of the image. This decision is made by considering the *'medpix_logo_bounding_box'* and *'vborder'* columns.
- **'lower crop'**: this is the location that has been selected to crop the bottom of the image. This decision is made by considering the *'hbar'* and *'hborder'* columns.
- **'visual_image_problems'**: this column contains the output of the convnet model, with the numbers following the words representing the probability that the image belongs to that category.
- **'invalid_image'**: this is a decision as to whether or not the image is invalid, e.g., has an arrow. This decision is made using the 'grayscale' and *'visual_image_problems'* columns as well as the text associated with the image (*'image_problems_from_text'*)
- **'invalid_image_reasons'**: in cases where the *'invalid_image'* column is True, column provides an account as to why a decision was made.

We can use this analysis to construct a new dataframe, with **'invalid_images'** removed and the remaining images cropped in such a way that problematic features are removed.

```
In [24]:
ip.clean_image_dataframe()
```

This 'cleaned' set, *should* have fewer instances of problematic images. One can check a random image from this new set:

```
In [25]:
#  show_image(ip.image_dataframe_cleaned['cleaned_
image'].iloc[180])
```

With time, the machinery used to detect these kinds of problems, particularly the convolutional neural network, will be improved. However, at the current time, **this class is still considered to be very experimental**.

7.5.5. Train, Validation and Test

Now that we've explored data harvesting, we can turn our attention to the final step before modeling: dividing data into training, validation and/or tests sets. Let's use images from the *Indiana University Chest X-Ray* collection* ('indiana_u_xray'). This set of images has been assembled 'by hand', and thus does not require complicated image cleaning procedures.

```
In [26]:
opi.search(collection='indiana_u_xray')
Results Found: 7,470.
```

Let's go ahead and download this entire collection.

Please be advised that this will take some time, so feel free to adjust download_limit to suit your needs.

```
In [27]:
pull_df2 = opi.pull(download_limit=None)
Number of Records to Download: 7,470 (chunk size:
30 records).
```

Let's quickly inspect this newly downloaded data.

```
In [28]:
simplify_df(opi.records_db_short)
Out[28]:
```

	mesh_major	mesh_minor	problems	abstract	affiliate	article_type	authors	cc_license	doc_source	image_caption	...	image_problems_from_text	imaging_modality_from_text	parsed_abstract	sex	modality_full	image_id_short	query	pull_time	cached_images_path	download_success
0	(Calcified Granuloma/lung/upper lobe/right,)		calcified granuloma	Comparison: Chest radiographs XXXX. Indicatio...	Indiana University	radiology_report	Kohli MD, Rosenman M	byncnd	CXR	PA and lateral chest x-XXXX XXXX.	...		Computed Tomography (CT): chest	{'impression': 'No acute cardiopulmonary proce...	male	X-Ray	1	{'subset': None, 'rankby': None, 'collection':...	2017-04-10 15:30:34.796070	/path/to/image	True
1	(Calcified Granuloma/lung/upper lobe/right,)		calcified granuloma	Comparison: Chest radiographs XXXX. Indicatio...	Indiana University	radiology_report	Kohli MD, Rosenman M	byncnd	CXR	PA and lateral chest x-XXXX XXXX.	...		Computed Tomography (CT): chest	{'impression': 'No acute cardiopulmonary proce...	male	X-Ray	2	{'subset': None, 'rankby': None, 'collection':...	2017-04-10 15:30:34.796070	/path/to/image	True
2	(normal,)		normal	Comparison: None. Indication: Positive TB tes...	Indiana University	radiology_report	Kohli MD, Rosenman M	byncnd	CXR	Xray Chest PA and Lateral	...		Computed Tomography (CT): chest	{'impression': 'Normal chest x-XXXX.', 'compar...		X-Ray	1	{'subset': None, 'rankby': None, 'collection':...	2017-04-10 15:30:34.796070	/path/to/image	True
3	(normal,)		normal	Comparison: None. Indication: Positive TB tes...	Indiana University	radiology_report	Kohli MD, Rosenman M	byncnd	CXR	Xray Chest PA and Lateral	...		Computed Tomography (CT): chest	{'impression': 'Normal chest x-XXXX.', 'compar...		X-Ray	2	{'subset': None, 'rankby': None, 'collection':...	2017-04-10 15:30:34.796070	/path/to/image	True
4	(Markings/lung/bilateral/interstitial/diffuse/...		markings; fibrosis	Comparison: None. Indication: dyspnea, subjec...	Indiana University	radiology_report	Kohli MD, Rosenman M	byncnd	CXR	CHEST 2V FRONTAL/LATERAL XXXX, XXXX XXXX PM	...		Computed Tomography (CT): chest	{'impression': 'Diffuse fibrosis. No visible f...		X-Ray	1	{'subset': None, 'rankby': None, 'collection':...	2017-04-10 15:30:34.796070	/path/to/image	True

5 rows × 37 columns

We can easily select a subset of these ~7000 images and divide them into training and test sets for some machine learning model using the image_divvy() tool.

```
In [29]:
from biovida.images import image_divvy
```

Let's imagine we're interested in building a model capable of distinguishing between 'normal' chest x-rays and those with signs of problematic calcium deposits, a disease formally known as 'calcinosis.'

We can define a rule to construct such a training and test set using a 'divvy_rule.' This rule will tell image_divvy() how to 'divvy up' the images in the cache. More specifically, our rule will tell this image_divvy() how to categorize images in the cache.

```
In [30]:
def my_divvy_rule(row):
    if isinstance(row['diagnosis'], str):
        if 'normal' in row['diagnosis']:
            return 'normal'  # though this could be
anything, e.g., 'super cool normal images.'
        elif 'calcinosis' in row['diagnosis']:
            return 'calcinosis'
```

Now that image_divvy() knows how we would like it to categorize the data we've downloaded, we can also pass it a dictionary specifying how to 'split' the data into training and testing sets. In this example, we'll use a standard 80% train, 20% test split and ask the function returns numpy arrays (ndarrays) as output.

```
In [31]:
train_test = image_divvy(instance=opi,
                          divvy_rule=my_divvy_rule,
                        db_to_extract='records_db',
                          action='ndarray',
                     train_val_test_dict={'train':
0.8, 'test': 0.2})
Structure:
- 'train':
  - 'calcinosis'
```

```
  - 'normal'
- 'test':
  - 'calcinosis'
  - 'normal'
```

Before signing off, image_divvy() printed the structure of the nested dictionary it returned.

We can use this information to unpack the arrays nested within this data structure:

```
In [32]:
train_ca, test_ca = train_test['train']['calcino-
sis'], train_test['test']['calcinosis']
train_norm, test_norm = train_test['train']['nor-
mal'], train_test['test']['normal']
```

Now that our data has been neatly unpacked, we can look at the number of samples the procedure generated.

```
In [33]:
# Normal
print("Train:", len(train_norm), "|", "Test:",
len(test_norm))
Train: 2156 | Test: 540
In [34]:
# Calcinosis
print("Train:", len(train_ca), "|", "Test:",
len(test_ca))
Train: 446 | Test: 112
```

Using the show_image() tool we imported above, we can take a quick at an image from each category.

```
In [35]:
# Normal
# show_image(train_norm[99])
In [36]:
# Calcinosis
# show_image(train_ca[104])
```

7.6. CONCLUSION

We have touched upon Biomedical Image Processing in this chapter using a Python package BioVida. I hope that Image Processing and drawing meaningful insights would become fun in days to come. All the best.

APPENDIX

APPENDIX 1

The goal of this appendix is to introduce working with the MIMIC-III database using Structured Query Language (SQL). While MIMIC-III is distributed as plain text files, it is easiest to work with the data within a relational database using SQL, partially due to the large number of concepts here which have been coded using SQL.

MIMIC-III is an openly available dataset developed by the MIT Lab for Computational Physiology, comprising deidentified health data associated with >60,000 hospital stays. Spanning 2001-2012, it includes demographics, vital signs, laboratory tests, medications, and more. A paper describing MIMIC-III is available from:

http://www.nature.com/articles/sdata201635

The dataset is provided as a collection of comma-separated value (CSV) files, which can be loaded into a database system such as PostgreSQL. A list of tables is provided on the MIMIC website:

http://mimic.physionet.org/mimictables/admissions/

We have highlighted some of the key tables below:

- *patients*: a list of patients covered in MIMIC-III, each identified by a unique subject_id.
- *admissions*: a list of hospital admissions, each identified by a unique hadm_id.

- *icustays*: a list of ICU stays, each identified by a unique icustay_id.

Where would I find a patient's date of birth?
The patient's date of birth can be found in the *patients* table, as detailed in the documentation:
http://mimic.physionet.org/mimictables/patients/
Where would I find a patient's hospital admission time?
The patient's hospital admission time can be found in the *admissions* table, which tracks hospital admission information, as detailed here:
http://mimic.physionet.org/mimictables/admissions/
Comma separated value (CSV) files are a plain text format used for storing data in a tabular, spreadsheet-style structure. While there is no hard and fast rule for structuring tabular data, it is usually considered good practice to include a header row, to list each variable in a separate column, and to list observations in rows.
As there is no official standard for the CSV format, the term is used somewhat loosely, which can often cause issues when seeking to load the data into a data analysis package. A general recommendation is to follow the definition for CSVs set out by the Internet Engineering Task Force in the RFC 4180 specification document.
Summarized briefly, RFC 4180 specifies that:

- files may optionally begin with a header row, with each field separated by a comma;
- records should be listed in subsequent rows. Fields should be separated by commas, and each row should be terminated with a line break;
- fields that contain numbers may be optionally enclosed within double quotes;
- fields that contain text ("strings") should be enclosed within double quotes;
- if a double quote appears inside a string of text then it must be escaped with a preceding double quote.

The CSV format is popular largely because of its simplicity and versatility. CSV files can be edited with a text editor, loaded as a spreadsheet in packages such as Microsoft Excel, and imported and processed by most data analysis packages. Often CSV files are an intermediate data format used to hold data that has been extracted from a relational database in preparation for analysis.

Relational Databases

Relational databases can be thought of as a collection of tables which are linked together by shared keys. Organizing data across tables can help to maintain data integrity and enable faster analysis and more efficient storage.

Motivation: why would we want a relational database?

Imagine trying to store data about a person: their name, age, and height. We can easily save this data in a CSV:

"Name", "Age", "Height"
"Penny", 30, 182

Now what if we measure Penny's heart rate every hour for four hours at 8:00am, 9:00am, 10:00am, and 11:00 am. How should we store this data? The naive approach would be to simply concatenate the information we have all in one file:

"Name", "Age", "Height", "Time", "Heart rate"
"Penny", 30, 182, "8:00", 65
"Penny", 30, 182, "9:00", 71
"Penny", 30, 182, "10:00", 72
"Penny", 30, 182, "11:00", 68

This works, but it feels very inefficient. We have repeated her name ("Penny"), her age (30), and her height (182) every time we get a heart rate measurement. The immediate solution is to not store both of these in the same file: we make one file for Penny's demographics (age, height), and we make another file for heart rate measurements. Then, we make sure that her name is the same in both, so that we know that both sets of data relate to Penny. We've created a relational database. Since her name is what links the two tables together, we would call the name column a "key."

Database Terminology

- "Database schema": The model that defines the structure and relationships of the tables.
- "Database query": Data is extracted from relational databases using structured "queries."
- "Primary key": A primary key is a field that uniquely identifies each row in a table.
- "Foreign key": A foreign key is a field that refers to a primary key in another table.

- "Normalization": The concept of structuring a database in a way that reduces data repetition and improves data integrity, usually by requiring one or more tables to be joined.
- "Renormalization": The concept of structuring a database to improve readability, sometimes at the expense of data repetition and data integrity.
- "Data type": A term used to describe the behavior of data and the possible values that it can hold (for example, integer, text, and date are all data types in PostgreSQL).

Giving a simple example of a hospital database with four tables, it might comprise of: a table with a list of all patients; a table with a log of hospital admissions; a table with a list of vital sign measurements; a table with a dictionary of vital sign codes and associated labels.

The patients table lists unique patients. The admissions table lists unique hospital admissions. The chart events table lists charted events such as heart rate measurements. The d_items table is a dictionary that lists itemids and associated labels, as shown in the example query. pk is primary key. fk is foreign key.

What is Structured Query Language (SQL)?

Structured Query Language (SQL) is a programming language used to manage relational databases. An SQL query has the following format:

```
SELECT [columns]
FROM [table_name];
```

The result of a query is generally a list of rows selected from your table/s of interest. For example, the following query lists the unique patient identifiers (subject_ids) of all female patients:

```
SELECT subject_id
FROM patients;
```

The * character is a wildcard that can be used to select all columns.

```
SELECT *
FROM patients;
```

WHERE keyword

Often you will want to select a subset of the data which satisfy some set of conditions. For example, you may want to select only female subjects from the database. This is easily accomplished with the WHERE keyword. The framework of our query becomes:

```
SELECT [columns]
FROM [table_name]
WHERE [conditions];
```

We can easily select all the subject_id corresponding to female subjects as follows:

```
SELECT subject_id
FROM patients
WHERE gender = 'F';
```

WHERE clauses are used to make a query return rows meeting only our specified criteria (our previous query, for example, returning only female patients). The simplest criteria is equality, WHERE gender = 'F.' Note that in this situation we specify a string, but this syntax will work for numbers as well. For example, we could select all the data for a single subject:

```
SELECT *
FROM patients
WHERE subject_id = 109;
```

WHERE clauses can be combined with standard logical operators AND/OR:

```
SELECT *
FROM patients
WHERE subject_id = 109
OR subject_id = 117
OR subject_id = 127;
```

A useful shorthand for OR statements on the same column is the IN condition:

```
SELECT *
FROM patients
WHERE subject_id IN (109, 117, 127);
```

We can also use the "less than" (<), "less than or equal to" <=, "greater than" (>), or "greater than or equal to" >= operators:

```
SELECT *
FROM patients
WHERE subject_id >= 109
```

```
AND subject_id <= 127;
```

SQL also offers shorthand for >= and <= combinations with the BETWEEN condition:

```
SELECT *
FROM patients
WHERE subject_id BETWEEN 109 AND 127;
```

Note the BETWEEN operator is inclusive. Verify for yourself that the above two queries give the same result.

When working with text data, we'll often want to search for partial string matches rather than exact matches. This can be accomplished with the LIKE keyword:

```
-- use `LIKE` to match text
-- The `%` is a wildcard that will match all characters
SELECT *
FROM icustays
WHERE first_careunit LIKE ‘%ICU%’;
```

Note the use of the wildcard character %.

ORDER BY keyword

The ORDER BY keyword is relatively straightforward: it will sort the data in the order you specify.

```
SELECT [columns]
FROM [table_name]
WHERE [conditions]
ORDER BY [columns];
```

The below query orders the results by the patient dob

```
SELECT subject_id, dob
FROM patients
ORDER BY dob;
```

Note that the WHERE clause is optional, and in the above query we have omitted it. However, we must respect the order of the keywords – and if we use the WHERE keyword it must appear after the FROM keyword and before the ORDER BY keyword.

Using SQL JOIN to query multiple tables

Often we need information coming from multiple tables. This can be achieved using the SQL JOIN keyword. There are several types of join, including INNER JOIN, LEFT JOIN, and RIGHT JOIN. It is important to understand the difference between these joins because their usage can significantly impact query results. Detailed guidance on joins is widely available on the web.

Using the INNER JOIN keyword, let's select a list of patients from the *patients* table along with dates of birth, and join to the *admissions* table to get the admission time for each hospital admission. We use the INNER JOIN to indicate that two or more tables should be combined based on a common attribute, which in our case is subject_id:

```
-- INNER JOIN will only return rows where subject_id
-- appears in both the patients table and the admissions table
SELECT p.subject_id, p.dob, a.hadm_id, a.admittime
FROM patients p
INNER JOIN admissions a
ON p.subject_id = a.subject_id
ORDER BY subject_id, hadm_id;
```

Performing operations on columns

Sometimes we will want to perform operations on columns. For example, if we are only interested in length of stay (los) to the nearest day, we can use the round function:

```
SELECT icustay_id, round(los)
FROM icustays;
```

Note that the column name ends up being round(los). We can specify the column name using the AS keyword:

```
SELECT icustay_id, round(los) AS los_integer_days
FROM icustays;
```

There are a large number of operations available in PostgreSQL.

Operations can involve multiple columns at once. For example, we may be interested in calculating, for patients who died, how long they spent in the hospital:

```
-- When combining columns in an operation, it is sometimes necessary
-- to convert ('cast') them to the same data type
SELECT subject_id, admittime, deathtime, deathtime – admittime AS length_of_stay
FROM admissions
WHERE deathtime IS NOT NULL;
```

Here we have introduced another concept: the IS NOT NULL clause which checks that the value is not null (a "null" is an empty value, and represents missing data).

How can we use temporary tables to help manage queries?

It is sometimes helpful to create temporary views or tables to break a large query into smaller, more manageable chunks. There are several approaches that can be used to create temporary tables. One method uses the WITH keyword. For example, we'll create a temporary view called patient_dates using the previous query, and then select all of its columns:

```
WITH patient_dates AS (
SELECT p.subject_id, p.dob, a.hadm_id, a.admittime,
  (cast(a.admittime as date) – cast(p.dob as date) / 365.2 ) as age
FROM patients p
INNER JOIN admissions a
ON p.subject_id = a.subject_id
ORDER BY subject_id, hadm_id
)
SELECT *
FROM patient_dates;
```

Another method is "materialized views", which create a new table on your database schema. We can then treat this view as any other database table.

```
-- we begin by dropping any existing views with the same name
DROP MATERIALIZED VIEW IF EXISTS patient_dates_view;
CREATE MATERIALIZED VIEW patient_dates_view AS
```

```
SELECT p.subject_id, p.dob, a.hadm_id, a.admittime,
  (cast(a.admittime as date) – cast(p.dob as date) / 365.2 ) as age
FROM patients p
INNER JOIN admissions a
ON p.subject_id = a.subject_id
ORDER BY subject_id, hadm_id;
```

CASE statement for if/else logic

The CASE statement is used to handle if/else logic. For example, using the *icustays* table you may want to group length of ICU stay (los) into short, medium, and long:

```
-- Use if/else logic to categorize length of stay
-- into ‘short’, ‘medium’, and ‘long’
SELECT subject_id, hadm_id, icustay_id, los,
  CASE WHEN los < 2 THEN ‘short’
      WHEN los >=2 AND los < 7 THEN ‘medium’
      WHEN los >=7 THEN ‘long’
      ELSE NULL END AS los_group
FROM icustays;
```

Aggregate functions

We are often interested in finding an aggregate value across multiple rows, such as the number of patients meeting a condition, an average heart rate, or a maximum blood pressure. We can do this using aggregate functions, such as COUNT(), MAX(), SUM(), and AVG().

Count the number of rows in the *icustays* table with the COUNT() function:

```
-- count the number of rows in a table
SELECT count(*)
FROM icustays;
```

Find the maximum length of stay in the *icustays* table with the MAX() function:

```
-- find the maximum length of stay in the ICU
```

```
SELECT MAX(los)
FROM icustays;
```

Aggregate are often combined with a GROUP BY clause, so that the aggregate function is applied to specific groups. For example, we can find the maximum length of stay, grouped for each patient:

```
-- find the maximum length of stay in the ICU
-- for each patient
SELECT subject_id, MAX(los)
FROM icustays
GROUP BY subject_id;
```

We may want to add a condition based on our new aggregate column. The WHERE clause won't filter on an aggregate column, so instead we use the HAVING keyword. For example, we can find the maximum length of stay, grouped for each patient, returning only patients whose maximum stay is less than 10 days:

```
-- find the maximum length of stay in the ICU
-- for each patient
-- where the maximum length of stay is < 10 days
SELECT subject_id, MAX(los)
FROM icustays
GROUP BY subject_id
HAVING MAX(los) <= 10;
```

Window functions

Sometimes an aggregate function isn't quite what we need. For example, we might want to create a column that lists the order of admissions to the ICU for each patient. In this case we do not want to group all of the rows with the same subject_id into a single row, so a simple a aggregate function is insufficient. Instead, we want to return multiple rows for each subject_id, with the order of admission computed over a subject_id window. For example, let's find the order of admission to the ICU for each patient using the RANK() window function:

```
-- find the order of admissions to the ICU for a patient
SELECT subject_id, icustay_id, intime,
```

```
  RANK() OVER (PARTITION BY subject_id ORDER BY intime)
FROM icustays;
```

We can then combine this query with a temporary table to select only the first ICU stay for each patient:

```
WITH icustayorder AS (
SELECT subject_id, icustay_id, intime,
  RANK() OVER (PARTITION BY subject_id ORDER BY intime)
FROM icustays
)
SELECT *
FROM icustayorder
WHERE rank = 1;
```

Multiple temporary views

Using the WITH statement, you can have more than one inline view. The services table contains information about what type of care a patient is receiving in the hospital.

```
-- find the care service provided to each hospital admission
SELECT subject_id, hadm_id, transfertime, prev_service, curr_service
FROM services;
```

Note that the services table doesn't have icustay_id, but we can join to it using hadm_id.

```
WITH serv as (
  SELECT subject_id, hadm_id, transfertime, prev_service, curr_service
  FROM services
)
, icu as
(
  SELECT subject_id, hadm_id, icustay_id, intime, outtime
  FROM icustays
)
```

```
SELECT icu.subject_id, icu.hadm_id, icu.icustay_id, icu.intime, icu.outtime
, serv.transfertime, serv.prev_service, serv.curr_service
FROM icu
INNER JOIN serv
ON icu.hadm_id = serv.hadm_id
```

However, something subtle has happened in this join. Let's see how many rows of data are returned by the above query. We can do this using the aggregate operator COUNT():

```
WITH serv as (
  SELECT subject_id, hadm_id, transfertime, prev_service, curr_service
  FROM services
)
, icu as
(
  SELECT subject_id, hadm_id, icustay_id, intime, outtime
  FROM icustays
)
SELECT COUNT(*)
FROM icu
INNER JOIN serv
ON icu.hadm_id = serv.hadm_id
```

Notice we have replaced all the column names with COUNT(*) – which means "count all the rows." What result do you get? Let's compare it to the original *icustays* table:

```
SELECT count(*)
FROM icustays;
```

Now you should be able to query the MIMIC-III database and take advantage of already created concepts – good luck!

APPENDIX 2

In this appendix we provide a procedure to setup a computational environment for tensorflow and keras.

Requirements

This tutorial requires the following packages:

- Python version 3.5
  - Python 3.4 should be fine as well
  - likely Python 2.7 would be also fine, but *who knows*? :P
- numpy version 1.10 or later: http://www.numpy.org/
- scipy version 0.16 or later: http://www.scipy.org/
- matplotlib version 1.4 or later: http://matplotlib.org/
- pandas version 0.16 or later: http://pandas.pydata.org
- scikit-learn version 0.15 or later: http://scikit-learn.org
- keras version 2.0 or later: http://keras.io
- tensorflow version 1.0 or later: https://www.tensorflow.org
- ipython/jupyter version 4.0 or later, with notebook support

(Optional but recommended):

- pyyaml
- hdf5 and h5py (required if you use model saving/loading functions in keras)
- **NVIDIA cuDNN** if you have NVIDIA GPUs on your machines. **Error! Hyperlink reference not valid.**

The easiest way to get (most) these is to use an all-in-one installer such as Anaconda from Continuum. These are available for multiple architectures.

Python Version

I'm currently running this appendix with **Python 3** on **Anaconda**

In [1]:

!python --version

Python 3.5.2

Configure Keras with tensorflow

1) Create the keras.json (if it does not exist):

touch $HOME/.keras/keras.json

2) Copy the following content into the file:

```
{
"epsilon": 1e-07,
"backend": "tensorflow",
"floatx": "float32",
"image_data_format": "channels_last"
}
```

In [2]:

```
!cat ~/.keras/keras.json
{
"epsilon": 1e-07,
"backend": "tensorflow",
"floatx": "float32",
"image_data_format": "channels_last"
}
```

Test if everything is up&running

1. Check import

In [3]:

```
import numpy as np
import scipy as sp
import pandas as pd
import matplotlib.pyplot as plt
import sklearn
```

In [4]:

```
import keras
```

Using TensorFlow backend.

2. Check installeded Versions

In [5]:

```
import numpy
print('numpy:', numpy.__version__)

import scipy
```

```
print('scipy:', scipy.__version__)

import matplotlib
print('matplotlib:', matplotlib.__version__)

import IPython
print('iPython:', IPython.__version__)

import sklearn
print('scikit-learn:', sklearn.__version__)
```

```
numpy: 1.11.1
scipy: 0.18.0
matplotlib: 1.5.2
iPython: 5.1.0
scikit-learn: 0.18
```

In [6]:

```
import keras
print('keras: ', keras.__version__)

# optional
import theano
print('Theano: ', theano.__version__)

import tensorflow as tf
print('Tensorflow: ', tf.__version__)
```

```
keras: 2.0.2
Theano: 0.9.0
Tensorflow: 1.0.1
```

If you have reached this point, with no errors then you are good to go. And welcome to the world of Deep Learning with Python ! Good Luck !

INDEX

D

E

F

G

H